THE SINGULARITY IN ANTIQUITY

The Singularity In Antiquity

AI AND THE ANCIENT MIND

GEW REPORTS & ANALYSES TEAM

Hichem Karoui (Ed.)

Global East-West (London)

Copyright © 2024 by by GEW REPORTS & ANALYSES TEAM
Hichem Karoui (Editor)
Global East-West (London)

All rights reserved. No part of this book may be reproduced in any manner whatsoever without written permission except in the case of brief quotations embodied in critical articles and reviews.

First Printing, 2024

Contents

1	Introduction: Bridging Millennia	1
2	Defining the Ancient Singularity	14
3	The Dawn of Ingenuity	26
4	Civilization's Cradle: Technology in Mesopotamia	40
5	Pyramids and Processing: AI in Ancient Egypt	53
6	Eastern Enigmas: Advanced Tech in Ancient China	65
7	The Technological Marvels of Ancient Persia	77
8	The Ingenious Innovations of Ancient Greece	90
9	Ancient Engineering in the Americas	103
10	The Mysteries of Mesoamerica	117
11	Automatons and Oracles: The Greek Insight	131
12	The Mechanical Wonders of Ancient Rome	144
13	The Legacy of Atlantis	157
14	Lost Lore and Mythical Machines	168
15	Uncovering Ancient Secrets	181
16	Deciphering the Ancients: Modern Interpretations	194
17	Implications for Today and Tomorrow	207
18	Aftermath: Reflections on the Ancient Singularity	220

19	The Future of Ancient Technology	232
20	Renaissance of Ancient Technology	245
21	Exploring Ancient Technologies in Science Fiction	258
22	Resurrecting Ancient Technologies	271
23	Ancient Wisdom in Modern Technology	285
24	Ancient Technology in Contemporary Art	298
25	Ancient Technology in Space Exploration	310
26	Legacy of Ancient Technology in the Modern World	323
27	Ancient Engineering in the Americas	336

1

Introduction: Bridging Millennia

Conceptual Prelude: Understanding Ancient Technology

Ancient technology acts as a prism through which we can glimpse the ingenuity and resourcefulness of our ancestors. To grasp the essence of ancient technology, one must recognize that it transcends mere artifacts and innovations; it embodies the collective intelligence and creative spirit of civilizations long past. As we delve into the theoretical foundations of ancient technology, we uncover a dynamic interplay of ideas, material culture, and societal needs. The term 'technology' itself is multifaceted in its historical context, encompassing not just mechanical devices but also systems of knowledge, techniques, and problem-solving approaches. In this exploration, we navigate the intricate web of concepts such as technological determinism, cultural evolution, and the diffusion of innovations. By understanding these ideas, we can comprehend how ancient societies harnessed their environments and crafted solutions to intricate challenges. This section aims to illuminate the rich tapestry of ancient technology, offering a comprehensive framework for examining its multifarious manifestations, from monumental architectural achievements to the everyday tools that shaped existence. Through this lens, we discern how ancient technology intertwined with social structures, religious practices, economic activities, and intellectual pursuits. Indeed, elucidating these connections empowers us to appreciate the underlying motivations and aspirations that fueled technological advancements in antiquity. Moreover, by conceptualizing ancient technology, we cultivate an enriched perspective on the scope and impact of human creativity across time. Ahistorical analysis enables us to contextualize the developments within their specific cultural milieus, shedding light on the diverse pathways taken by different civilizations while illustrating the common threads of innovation that bind humanity together. Additionally, our exploration encompasses the philosophical underpinnings of technological progress, dissecting doctrines such as teleology, pragmatism, and the relationship between science and technology. We interrogate the triumphs and tribulations of ancient innovators, scrutinizing the ethical dimensions of their creations and endeavors. Ultimately, delving into the conceptual nuances of ancient technology fosters a nuanced appreciation for the interwoven tapestry of human history and provides profound insights into the intricacies of our shared heritage.

Scope and Intentions of the Book

The scope and intentions of this book delve into the depths of ancient technological advances, exploring how innovation from millennia ago continues to reverberate through contemporary society. By elucidating the remarkable achievements of our ancestors, this work aims to bridge temporal gaps and provide a comprehensive understanding of the evolution of technology. In doing so, we seek to illuminate how ancient knowledge can inform and inspire modern endeavors, fostering a deeper appreciation for the ingenuity of early civilizations. Through a thorough examination of historical artifacts, texts, and archaeological findings, we endeavor to shed light on the enduring legacy of ancient technologies. Our intention is not merely to catalog these innovations, but to underscore their relevance in shaping the trajectory of human progress, steering the course of scientific discovery, engineering ingenuity, and societal development. This book also seeks to challenge conventional views of the past and prompt readers to contemplate the interconnectedness of history and technological advancement. Moreover, by highlighting the resilience and adaptability of ancient technologies, we hope to spark dialogue around sustainable practices and innovative solutions for contemporary challenges. Ultimately, the ambition of this book is to encourage readers to envision a future that draws on the timeless wisdom of our predecessors while embracing the possibilities of cutting-edge technology. Throughout the chapters that follow, we will embark on a profound expedition into the annals of history, unveiling the profound significance of ancient technologies and their enduring impact on the contemporary world.

Technological Continuity Through Ages

Throughout the annals of history, the evolution of technology has woven a tapestry that spans the millennia, connecting disparate cultures and civilizations across time. This interconnectedness manifests as an intricate web of innovation, with each generation building upon the knowledge and advancements of its predecessors. In exploring the concept of technological continuity through the ages, we are compelled to delve into the profound interplay between ancient ingenuity and modern accomplishments. From rudimentary tools and simple machines to the complexities of artificial intelligence and nanotechnology, the persistent thread of advancement connects us to the distant past, reaffirming our common heritage as creators and problem-solvers. The enduring spirit of innovation has transcended cultural boundaries and geographical limitations, leaving an indelible mark on the collective human narrative. The notion of continuity encompasses not only the technical aspects of ancient innovations but also the underlying principles and ideologies that have stood the test of time. Through the transmission of knowledge and the preservation of intellectual legacies, our understanding of the interconnected nature of technological developments becomes more profound. The remarkable similarities found in the architectural precision of ancient monuments, the sophisticated water management systems, and the celestial navigation tools serve as compelling evidence of an unbroken lineage of technological prowess. As we seek to comprehend the intricacies of ancient technological achievements, we are confronted with the realization that our contemporary advancements are rooted in the enduring legacy of our forebears. Furthermore, the intergenerational exchange of ideas and techniques has fostered a dynamic continuum of progress, propelling humanity forward despite the ebbs and flows of history. A striking illustration of this continuity can be found in the domains of medical knowledge, where ancient remedies and surgical practices laid the groundwork for modern healthcare systems. Similarly, the advances in agricultural methods, metallurgy, and transportation bear witness to the enduring impact of ancestral wisdom on our present-day pursuits. By tracing the trajectories of technological evolution, we gain invaluable insights into the perpetual fusion of the old and the new, fostering an appreciation for the timeless relevance of innovation. The profound implications of this continuity resonate far beyond the realm of scientific inquiry, extending into the realms of cultural identity, societal values, and ethical considerations. By recognizing the shared lineage of technological achievements, we are invited to reevaluate our perspectives on progress and development, discerning the ethical imperatives embedded within the pursuit of innovation. As we navigate the intersection of tradition and transformation, we are compelled to embrace a holistic approach to technological evolution that honors both historical precedents and future aspirations. Embracing the nuanced interplay between the ancient and the contemporary, we lay the foundation for a more inclusive and enlightened discourse on the trajectory of human progress.

Historical Relevance and Modern Implications

The study of ancient technology brings to light the profound historical relevance and modern implications that continue to shape our understanding of innovation and progress. By delving into the technological achievements of our predecessors, we gain valuable insights into the evolution of human civilization and the enduring impact of ancient ingenuity on contemporary advancements. Exploring the developments of antiquity allows us to discern patterns, anticipate future possibilities, and appreciate the significant role that technological innovation has played in shaping human societies throughout history. As we examine the historical relevance of ancient technology, we are confronted with the realization that many foundational concepts and inventions from the past continue to underpin our modern technological landscape. Whether it be the early mathematical innovations of ancient Mesopotamia, the sophisticated hydraulic engineering of ancient Rome, or the impressive astronomical knowledge of ancient civilizations, the enduring influence of these advancements is unmistakable. Such historical continuity serves as a testament to the enduring legacy of ancient technologies, which have left an indelible mark on our contemporary world. Furthermore, recognizing the historical relevance of ancient technology enables us to draw parallels to our current technological era, as we navigate the complexities of rapid digital transformation, artificial intelligence, and sustainable engineering. By understanding the historical contexts of technological breakthroughs, we can better appreciate the challenges and opportunities inherent in our own epoch, identifying potential pitfalls, and striving for ethical, sustainable, and impactful innovation. The modern implications of ancient technology extend far beyond mere historical curiosity, offering substantive lessons for how we approach innovation, problem-solving, and societal progress today. In studying the methods, materials, and ideologies behind ancient technologies, we glean valuable perspectives on resilience, resourcefulness, and adaptability that remain relevant in our rapidly changing world. Furthermore, by acknowledging the historical resonance of ancient artifacts and designs, we are prompted to consider the ethical and environmental implications of our current technological pursuits, fostering a deeper awareness of our responsibilities as stewards of innovation and progress. Ultimately, the historical relevance and modern implications of ancient technology furnish us with a comprehensive framework from which to evaluate and appreciate the rich tapestry of human ingenuity over millennia. By examining the enduring legacies of ancient innovations, we can derive critical insights, inspiration, and wisdom that not only enrich our understanding of history but also inform our contributions to the ongoing narrative of technological advancement and societal development.

Methodology and Sources

The exploration of ancient technology demands a comprehensive and meticulous approach to research and analysis. Given the scarcity of direct evidence from millennia-long bygone eras, deriving an understanding of ancient technologies relies heavily on an interdisciplinary methodology that draws upon diverse sources and analytical tools. This chapter expounds upon the rigorous processes and varied sources that inform our examination of ancient technological marvels. At the core of our methodology lies the critical assessment of archaeological findings, textual records, and iconographic depictions. Archaeological excavations offer invaluable insights into early material culture, providing tangible artifacts and physical remains that serve as tangible evidence of ancient technological advancements. Additionally, deciphering and interpreting ancient texts, including cuneiform tablets, papyri, and inscriptions, aids in unraveling the intricacies of ancient engineering and innovation. Iconographic representations on pottery, murals, and reliefs also constitute vital sources, offering visual narratives of ancient technologies and societal applications. Furthermore, this endeavor necessitates a synthesis of historical and anthropological perspectives to construct a holistic understanding of ancient technology within its socio-cultural contexts. By delving into ancient societies' belief systems, economic structures, and geopolitical dynamics, we can discern the motivations and implications underlying technological advancements. In conjunction with traditional sources, scientific analyses play a pivotal role in affirming and broadening our comprehension of ancient technology. Cutting-edge techniques such as radiocarbon dating, isotopic analysis, remote sensing, and experimental archaeology complement the study of ancient artifacts and architectural remnants, facilitating the reconstruction of technological processes and validating historical narratives. To ensure the credibility and objectivity of our interpretation, it is imperative to critically evaluate the reliability and biases inherent in our sources. The nuanced interplay between primary, secondary, and tertiary sources mandates a rigorous vetting process to discern conjecture from authenticated information, thereby upholding scholarly integrity and accuracy. By synthesizing these multifaceted approaches, this book endeavors to present a comprehensive and rigorously researched exploration of ancient technology, providing readers with a lucid and informed perspective on the innovations that bridged epochs.

Challenges in Interpreting Ancient Innovations

Interpreting ancient innovations presents a myriad of challenges that require a delicate balance of scholarly rigor and creativity. The primary hurdle lies in the scarcity of direct evidence, as much of the technology from antiquity has either vanished or left behind only fragmentary remains. This necessitates an interdisciplinary approach that draws on archaeology, history, anthropology, and other fields to piece together a cohesive understanding of ancient technology. Another critical challenge is deciphering the original intentions and functions of these innovations. Without firsthand accounts, we must rely on conjecture, comparative analysis, and experimentation to extrapolate the purpose and utility of ancient devices. Additionally, the cultural context in which these technologies emerged adds layers of complexity, as our interpretations must be attuned to the values, beliefs, and societal structures of bygone civilizations. Furthermore, the limitations of modern language present significant obstacles when attempting to articulate and comprehend the intricacies of ancient technology. Technical terminology, which varies across cultures and epochs, often defies direct translation or analogy, necessitating careful contextualization and nuanced communication. The preservation and authenticity of artifacts pose yet another challenge, as provenance, restoration, and the potential for tampering or misinterpretation can cast doubt on the integrity of our insights. It is vital to navigate these uncertainties with meticulous attention to detail and a discerning eye for historical accuracy. Moreover, the influence of preconceived notions and contemporary biases can inadvertently distort our perceptions of ancient technological achievements. Avoiding anachronistic projections requires a conscious effort to adopt a mindset that transcends present-day paradigms, fostering an impartial lens with which to view the past. Ultimately, the quest to interpret ancient innovations demands an embrace of ambiguity while striving for clarity. The process involves constant refinement, collaborative discourse, and a recognition of the inherent limitations of reconstructing the technologically rich tapestry of antiquity.

Significance of Bridging Temporal Gaps

The quest to bridge temporal gaps in understanding ancient technologies is much more than a mere academic pursuit. It is a fundamental endeavor that can significantly enrich our comprehension of human history, the evolution of civilization, and the development of technology over millennia. By delving into the depths of the past, we aim to unravel the intricate tapestry of human ingenuity, creativity, and problem-solving across diverse cultures and eras. Indeed, by connecting the dots between ancient and modern technologies, not only do we gain a deep appreciation for the intellectual prowess of our predecessors, but we also glean invaluable insights that can inform contemporary innovation and technological advancements. Furthermore, bridging these temporal gaps offers a unique opportunity to explore the interconnectedness of human societies throughout history. It allows us to identify recurring patterns of innovation, diffusion of knowledge, and the enduring impact of technological achievements on shaping human experiences. Through this lens, we can recognize the universal human quest for knowledge and progress, transcending temporal and geographical boundaries. Understanding how ancient civilizations harnessed natural resources, applied scientific principles, and devised ingenious solutions to overcome challenges fosters a sense of shared human heritage and resilience, fostering cross-cultural empathy and mutual appreciation. Moreover, uncovering and comprehending ancient technologies can have profound implications for contemporary society. It provides a historical context for evaluating the current trajectory of technological development and prompts critical reflections on ethical, environmental, and societal aspects of innovation. By scrutinizing the outcomes of ancient technological endeavors, we are prompted to reassess our approaches to technology, emphasizing sustainable, inclusive, and ethically grounded practices. This retrospective analysis offers a valuable perspective for addressing the complex issues and dilemmas posed by cutting-edge technologies, offering guidance for steering present-day innovations toward responsible and beneficial outcomes. In essence, bridging temporal gaps in understanding ancient technologies serves as a conduit for cultivating a holistic view of human heritage, opening avenues for cross-disciplinary dialogue, and laying the foundations for informed and conscientious technological progress. As we embark on this exploratory journey through the annals of history, we endeavor to unearth the inherent wisdom and timeless lessons encapsulated within the artifacts and innovations of antiquity, ultimately enriching our collective narrative and guiding our endeavors towards a more enlightened future.

Preview of Significant Discoveries

In delving into the rich tapestry of human history, we uncover a trove of remarkable technological achievements that have often been overlooked or understated. By examining these ancient innovations through a contemporary lens, we offer our readers an illuminating glimpse into the astonishing breadth of human ingenuity across the millennia. This exploration spans continents and civilizations, encapsulating the sheer diversity and brilliance of early technological feats. From the astounding engineering marvels of the ancient world to the enigmatic remnants of advanced knowledge, our journey will bring to light the ingenious creations that continue to defy easy explanation. We will traverse the majestic pyramids of Egypt, each stone a testament to unparalleled craftsmanship and architectural precision. We will unravel the cryptic allure of the Antikythera Mechanism, an ancient Greek device that stands as a beacon of early computational prowess. Our expedition will lead us through the sophisticated irrigation systems of Mesopotamia, revealing the foundation upon which modern agriculture and water management were built. Furthermore, we shall uncover the enigmatic legends surrounding the lost city of Atlantis, contemplating the tantalizing possibility of a highly advanced civilization existing in antiquity. Each chapter unfolds a new narrative, woven from the threads of discovery and speculation. As we embark on this voyage, we invite our readers to embrace the mysteries of the past and to witness the enduring legacy of ancient innovation that continues to shape our world today. Embracing the technological marvels of our forebears allows us to cultivate a profound appreciation for the depth of human achievement, while simultaneously kindling the flames of curiosity and imagination. It is our hope that by peering into the annals of antiquity, readers will find inspiration in the resilience and resourcefulness of our ancestors, fostering a deeper understanding of the ever-evolving relationship between humanity and technology.

The Journey Ahead: Summary of Subsequent Chapters

In the chapters to come, readers will embark on an extraordinary exploration of ancient technological advancements that have transcended time. We will delve into the origins of innovation, examining early technological marvels that paved the way for future ingenuity. The narrative will then evolve to explore the concept of an ancient singularity, delving deep into a period of exponential growth and transformative technological leaps that occurred in various civilizations across different continents. Our journey will traverse through Mesopotamia, Egypt, China, Persia, Greece, Mesoamerica, and the Indus Valley, uncovering the remarkable technological achievements of these ancient societies. Through meticulous research and analysis, we will unravel the mysteries surrounding advanced tech applications, artificial intelligence, automation, and mechanical innovations that once flourished in these civilizations. As we progress, our discussion will extend to the implications of these ancient technologies on contemporary society, offering thought-provoking insights into their relevance and potential integration with modern advancements. Furthermore, the subsequent chapters will shed light on the influence of ancient technology on science fiction, contemporary art, space exploration, and its relation to the Renaissance of ancient wisdom in the present era. Through this comprehensive and captivating journey, readers will gain a profound understanding of the enduring impact of ancient technology on our world today, and the possibilities it presents for the future. Drawing inspiration from the boundless creativity and ingenuity of our predecessors, we will cultivate a deep appreciation for the timeless legacy of ancient technological innovations, igniting aspirations for innovative endeavors that build upon the rich tapestry of human achievement.

Inspiration and Aspirations

The study of ancient technology is more than just an exploration of the past; it is a testament to the human spirit's enduring quest for knowledge and innovation. The rich tapestry of historical achievements that we have uncovered speaks to the resilience and indomitable curiosity of our forebears. As we embark on this intellectual odyssey, it is essential to recognize the profound inspiration that emanates from the technological marvels of antiquity. These timeless creations, born from the ingenuity of ancient civilizations, serve as beacons of aspiration, guiding us on a voyage through time and offering invaluable lessons that resonate with contemporary society. One of the key aspirations of delving into the realms of ancient technology is to rekindle a sense of wonder and admiration for the ingenious accomplishments of our predecessors. By studying the innovative solutions devised by ancient societies, we gain a deeper appreciation for the inventive prowess that transcends age and era. Witnessing the mechanical marvels and architectural wonders of ancient times ignites a spark within our own creative faculties, compelling us to envision new possibilities and push the boundaries of modern craftsmanship and engineering. Furthermore, the exploration of ancient technology kindles a desire to bridge the gap between bygone eras and the present day. It fosters an understanding of the enduring relevance of technological insights from millennia ago, highlighting the interconnectedness of human achievement across time and space. This endeavor embodies a yearning to connect with our heritage, honor the legacy of our ancestors, and weave a narrative of continuity that transcends temporal confines. Through this pursuit, we aspire to imbue our contemporary undertakings with the wisdom and sagacity distilled from the annals of history, ensuring that our technological advancements are imbued with the time-tested principles and innovations that have withstood the test of time. Ultimately, the aspiration underlying this journey into ancient technology encompasses a symbiotic blend of reverence for the past and vision for the future. By drawing sustenance from the collective wisdom of antiquity, we strive to chart a path toward progressive innovation and sustainable advancement. Our aspirations extend beyond mere theoretical inquiry; they encompass a larger ambition to harness the enduring legacy of ancient technology as a catalyst for constructive change in modern society.

Through our endeavors, we seek to inspire a renaissance of creativity, foster cross-cultural collaboration, and pave the way for a harmonious integration of tradition and innovation. This ambitious pursuit holds the promise of unlocking novel perspectives, rejuvenating interest in historical scholarship, and catalyzing transformative developments that resonate across disciplines and generations. In this light, our overarching aspiration is to invoke a renewed appreciation for the profound cultural, scientific, and philosophical heritage encapsulated within ancient technology and leverage this appreciation as a springboard for constructive progress in the contemporary world.

2

Defining the Ancient Singularity

Introduction to the Concept of Singularity

The concept of singularity, in the context of technological advancement, represents a theoretical point where progress accelerates beyond human comprehension and control. It encapsulates the idea of an event horizon in time, marking a threshold of unprecedented transformation propelled by the exponential growth of knowledge and innovation. Singularities, as theoretical constructs, serve as focal points for contemplating the limits of human ingenuity and the potential emergence of transcendent technologies. This enigmatic concept has deep philosophical roots, often intertwined with discussions on artificial intelligence, post-humanism, and speculative futurism. Within the domain of ancient technologies, the notion of singularities unveils intriguing prospects for reevaluating historical achievements. By framing ancient innovations through the lens of singularity theory, it becomes possible to discern patterns of rapid technological development and identify instances where civilizations may have approached or even crossed conceptual thresholds akin to modern-day singularities. This approach not only enriches our understanding of ancient societies but also instigates reflections on the cyclical nature of technological evolution and the recurring motifs of intellectual leaps throughout human history.

Historical Perspectives on Singularities

The concept of a technological singularity, a hypothetical future event in which artificial intelligence or other technological advancements trigger runaway technological growth, has garnered significant attention in modern times. However, historical perspectives suggest that the notion of a 'singularity' is not exclusive to the contemporary era. Ancient civilizations evidenced remarkable advancements that could be viewed through the lens of technological singularities.

Looking back through history, we can identify pivotal moments when human societies experienced transformative leaps in technological capabilities. These paradigm shifts, comparable to the modern conception of a singularity, reshaped the course of civilization and fostered unprecedented advancements. For instance, the development of writing systems in Mesopotamia and Egypt marked a revolutionary leap in communication and record-keeping, setting the stage for subsequent intellectual and technological progress.

Ancient texts and artifacts also serve as compelling evidence of singularities in the past. The Antikythera Mechanism, an ancient Greek analog computer, provides a striking example of an advanced technological achievement that defies conventional timelines of technological progression. Similarly, the marvels of engineering evidenced in the construction of the Egyptian pyramids and the precision of astronomical knowledge possessed by ancient cultures raise intriguing questions about the existence of ancient technological singularities.

Furthermore, the diffusion of knowledge and technological innovations across ancient empires and civilizations underscores the potential interconnectedness of historical singularities. Trade routes, diplomatic exchanges, and conquests facilitated the exchange of ideas, practices, and technologies, contributing to the dissemination of groundbreaking advancements beyond regional boundaries. This dissemination of knowledge and technology may have precipitated widespread technological acceleration akin to the characteristics of singularities.

Considering historical evidence through the framework of technological singularities encourages a reevaluation of our understanding of past civilizations and their technological capabilities. It prompts us to explore the possibility of surges in innovation and knowledge that propelled ancient societies into eras of unparalleled advancement, laying the groundwork for subsequent developments. By delving into historical perspectives on singularities, we gain valuable insights into the dynamics of technological evolution and its enduring impact on humanity.

Criteria for Identifying Ancient Singularities

Ancient singularities possess distinct characteristics that differentiate them from periods of technological progression. When seeking to identify ancient singularities, it is essential to establish specific criteria and parameters that can be applied universally across various civilizations and historical contexts. One crucial criterion is the rapid and unprecedented advancement in multiple areas of technology and knowledge within a relatively short timeframe. This rapid acceleration sets ancient singularities apart from typical development patterns, indicating an extraordinary and potentially transformative event. Additionally, these advancements should demonstrate a profound impact on the society, economy, culture, and overall trajectory of the civilization in question. The innovative breakthroughs should not only revolutionize existing practices but also lay the groundwork for substantial future developments. Furthermore, the presence of interconnected and synergistic technological advancements across different fields, such as engineering, astronomy, mathematics, and medicine, serves as a key indicator of an ancient singularity. These intertwined advancements collectively contribute to an exponential growth in overall knowledge and capabilities, signifying a pivotal juncture in human history. It is imperative to consider the sustainability and longevity of these advancements when evaluating ancient singularities. Sustainable implementations that endure beyond the initial phase of innovation and continue to influence subsequent generations underline the magnitude of the technological leap. Moreover, the enduring impact of these innovations on global or regional scales provides valuable insight into the singularity's significance within a broader historical context. Lastly, the impact on philosophical, religious, and cultural paradigms should not be overlooked, as ancient singularities often trigger profound shifts in belief systems, cosmology, and societal structures, reflecting the far-reaching implications of technological and intellectual advancements. By applying these criteria, scholars and historians can systematically identify and evaluate ancient singularities, shedding light on the remarkable episodes of innovation that have shaped the course of human civilization.

Technological Thresholds and Civilization

Technological innovation has been a defining factor in the development of human civilization, marking significant thresholds that have propelled societies forward. It is imperative to recognize the pivotal role that technology plays in shaping the course of human history, specifically in relation to ancient civilizations. At various junctures, technological advancements have served as catalysts for societal transformations, influencing cultural, economic, and political landscapes.

The emergence of new technologies often signifies a shift in the capabilities and aspirations of a civilization. From the mastery of fire to the development of agricultural techniques, each breakthrough has had profound implications. These milestones not only reflect the adaptability and ingenuity of human communities but also signify a progression towards more complex and cohesive social structures.

Moreover, technological thresholds have fostered interconnectedness and communication among diverse populations, thereby contributing to the exchange of knowledge and ideas. The invention of written language, for instance, expanded the boundaries of human expression and facilitated the documentation of accumulated wisdom and discoveries. This exchange of knowledge formed the foundation for subsequent innovations, creating a continuous cycle of advancement that defined the trajectory of early civilizations.

Furthermore, the attainment of technological thresholds engendered shifts in power dynamics within societies. The development of advanced tools and machinery often led to the rise of skilled labor groups and specialized professions, subsequently shaping hierarchical structures. These reconfigurations laid the groundwork for organized systems of governance, reinforcing the symbiotic relationship between technology and societal structure.

As civilizations evolved and encountered new challenges, the quest for innovation became intrinsic to their survival and progress. Technological thresholds prompted intricate problem-solving approaches, fostering scientific inquiry and experimentation. This pursuit of understanding natural phenomena and harnessing their potential epitomized the intellectual growth of early civilizations and set the stage for future scientific endeavors.

In essence, the recognition of technological thresholds as pivotal markers in the narrative of human civilization underscores the dynamic relationship between innovation and societal progression. As we delve into the examination of ancient singularities, it is imperative to contextualize the significance of technological thresholds in elucidating the evolutionary path of ancient civilizations.

Case Studies: Instances of Ancient Singularity

The concept of ancient singularity is not merely a theoretical construct but finds its basis in the tangible evidence provided by historical, archaeological, and technological discoveries. In this section, we will delve into fascinating case studies that illustrate instances of profound technological advancement in ancient civilizations. These case studies will serve as compelling examples of the existence of ancient singularities, offering insights into the remarkable achievements of our ancestors.

One such remarkable case study takes us to the Hellenistic city of Alexandria, where a convergence of extraordinary advancements in various fields occurred during the reign of the Ptolemaic dynasty. The construction of the Great Library of Alexandria, which housed an extensive collection of knowledge from across the ancient world, stands as a testament to the intellectual and technological prowess of the era. The library became a focal point for scholars, inventors, and scientists, fostering an environment that catalyzed innovations in astronomy, mathematics, medicine, and engineering.

Another compelling case study leads us to ancient China, specifically during the Han Dynasty. This period witnessed significant advancements, including the invention of paper, the development of intricate irrigation systems, and remarkable progress in metallurgy and ceramics. The construction of the Grand Canal, an engineering marvel that facilitated transportation and trade, exemplifies the sophistication of ancient Chinese technological endeavors. These achievements signify the existence of a technological singularity within the context of ancient China, marking a period of rapid advancement and innovation.

Moving to the Americas, the enigmatic civilization of the Maya presents yet another compelling case study. The precision and complexity of their architectural achievements, astronomical knowledge, and calendrical systems defy conventional understanding and stand as a testament to the sophisticated technological prowess of the Maya civilization. The intricately designed city-states, advanced agricultural practices, and intricate hieroglyphic script showcase a level of technological development that aligns with the attributes of an ancient singularity.

Furthermore, the Indus Valley Civilization offers intriguing case studies that signal the presence of ancient technological singularity. The well-planned urban centers and complex drainage systems reflect an advanced understanding of urban planning and hydraulic engineering that surpasses contemporary developments in other regions of the ancient world. Additionally, the sophisticated craftsmanship displayed in artifacts and the layout of cities such as Mohenjo-Daro and Harappa hint at an era of technological acceleration and innovation.

These diverse case studies not only provide compelling evidence for ancient singularities but also prompt reevaluations of our perceptions of ancient civilizations. By exploring these instances of exceptional technological accomplishments, we gain a deeper appreciation for the ingenuity and innovation that characterized the ancient world, paving the way for a more comprehensive understanding of the concept of ancient singularity.

Analyzing Causes of Technological Acceleration

Technological acceleration in ancient civilizations was a phenomenon that has intrigued historians, archaeologists, and technologists for centuries. The rapid development and deployment of innovative technologies in various ancient societies raise important questions about the causes behind such acceleration. To analyze these causes, it is necessary to delve into the specific factors that facilitated technological advancement during these ancient periods. One key factor is the availability of resources and materials. Access to abundant natural resources such as metals, minerals, and fertile land often spurred technological innovation and progress. Moreover, the presence of skilled artisans, engineers, and craftspersons played a pivotal role in accelerating technological development. Their expertise and creativity enabled the conception and realization of groundbreaking inventions and tools. Another crucial element is the exchange of knowledge and ideas through trade networks and cultural interactions. Cross-cultural exchanges facilitated the transfer of technological know-how, leading to the adoption and adaptation of new concepts and techniques. Additionally, the influence of societal needs cannot be overlooked. Necessity often became the mother of invention, prompting ancient societies to innovate in response to challenges such as warfare, agriculture, and commerce. Moreover, the role of patronage from rulers, nobility, and religious institutions cannot be understated. Financial and moral support provided by these influential figures encouraged the pursuit of technological endeavors, motivating individuals and collectives to pursue ambitious projects. Furthermore, environmental and climatic factors also contributed to technological acceleration. Adaptation to changing environmental conditions, such as climate shifts or natural disasters, necessitated the development of resilient and efficient technologies. Finally, the presence of established educational and scholarly systems promoted the accumulation and dissemination of knowledge, fostering an environment conducive to technological progress. By scrutinizing these multifaceted causes, we can gain a deeper understanding of the complex forces that drove technological acceleration in ancient civilizations and draw valuable insights for contemporary technological development.

Impact on Society and Culture

The concept of ancient technological singularity has far-reaching implications for the understanding of societal and cultural developments in ancient civilizations. As technology advanced exponentially, it would have had a profound impact on the fabric of society and the expression of culture. One key area of impact is the reconfiguration of power structures within ancient societies. The emergence of advanced technologies could have led to new hierarchies and systems of authority, with those possessing or controlling innovative technologies gaining significant influence. This could have influenced social dynamics, governance structures, and interactions between different segments of the population. Moreover, the integration of sophisticated technologies into daily life would have influenced cultural practices and belief systems. It is likely that the presentation of remarkable technological feats would have been intertwined with religious or mythological narratives, shaping ideologies and belief systems. Additionally, the proliferation of advanced tools and machinery might have transformed labor practices and economic systems, leading to shifts in occupational roles and trade networks. Furthermore, the adoption of cutting-edge inventions could have influenced artistic expressions, architectural achievements, and intellectual pursuits, resulting in a renaissance of creativity and innovation. Another important aspect to consider is the potential impact on communication and the dissemination of knowledge. Advanced technologies would have altered the means and speed by which information was shared, potentially contributing to the spread of ideas, languages, and cultural exchanges. The societal and cultural impacts of an ancient technological singularity are complex and multifaceted, encompassing domains such as governance, religion, economy, arts, and knowledge transmission. Understanding these impacts provides valuable insights into the dynamics of ancient civilizations and their enduring influence on contemporary societies.

Comparative Analysis with Modern Singularity Theories

The concept of singularity, whether ancient or modern, remains a topic of profound intellectual debate and speculative conjecture. As we delve deeper into the comparative analysis between ancient and modern singularity theories, it is essential to examine both the divergences and convergences that exist between the two. In modern theoretical frameworks, singularity often denotes the notion of an unprecedented acceleration of technological development leading to a profound transformation of human civilization. In contrast, when contemplating ancient singularities, the focus shifts towards discerning historical instances where technological advancements or societal changes seemed to exceed the anticipated pace of progress. Drawing parallels between these distinct orientations presents a fascinating challenge. One avenue for comparative analysis involves investigating the catalysts behind each type of singularity. While modern singularity theories often center around exponential growth in computing power and artificial intelligence, ancient singularities may have been propelled by other pivotal innovations such as the invention of writing systems, advanced construction techniques, or remarkable philosophical insights. Moreover, the impact on society and culture, a fundamental element within modern singularity discussions, provides a compelling reference point for comparison. Examining the societal repercussions of ancient technological leaps offers valuable insight into how such transformations affected ancient communities, shaping their belief systems, social structures, and economic dynamics. Additionally, scrutinizing the aftermath of ancient singularities and comparing them to the expected consequences of modern theoretical singularities can shed light on the potential long-term ramifications of rapid technological shifts. Furthermore, engaging in a cross-temporal comparative analysis allows for a dialectical exploration of the pitfalls and strengths inherent in both ancient and modern singularity perspectives. While critiques of contemporary singularity theories raise concerns about ethical implications, AI dominance, and existential risks, critiques of ancient singularity hypotheses might highlight the challenges of interpreting fragmentary historical evidence or attributing causality. Integrating archaeological, historical, and philosophical methods with insights from contemporary futurism and technological forecasting can enrich our understanding of this crossroads between past and future. Through this comparative analysis, we aim to discern overarching patterns that transcend temporal boundaries, fostering a comprehensive grasp of the complex relationship between human innovation and societal transformation.

Criticisms and Controversies

The concept of the Ancient Singularity has sparked significant debate and controversy within scholarly and scientific communities. One of the key criticisms revolves around the fallibility of identifying instances of ancient technological revolutions and determining their impact. Skeptics argue that attributing advanced technological progress to specific ancient civilizations may oversimplify complex historical processes and neglect the broader socio-cultural context in which these developments occurred. Additionally, the lack of empirical evidence and specific criteria for identifying ancient singularities raises concerns about the validity and objectivity of such claims. Critics caution against imposing modern interpretations and theories onto ancient societies, emphasizing the importance of nuanced historical analysis and interdisciplinary collaboration. Furthermore, controversies arise from differing perspectives on the nature of technological acceleration in premodern civilizations. Some scholars argue that the notion of a singularity is inherently tied to the trajectory of industrialization and the Information Age, and applying it to ancient contexts may be anachronistic. They contend that while remarkable technological achievements are evident in antiquity, these do not necessarily align with the exponential growth and transformative impact characteristic of modern technological revolutions. Consequently, the criteria for defining an ancient singularity remain a subject of considerable disagreement and deliberation. Moreover, ethical and cultural considerations contribute to the contentious nature of the discourse surrounding ancient technological advancements. Debates regarding the portrayal of ancient cultures as precursors to modern innovation raise questions about cultural appropriation, historical agency, and the potential distortion of diverse cultural narratives. Criticisms also surface in relation to the implications of attributing exceptional technological accomplishments to ancient civilizations, particularly concerning issues of exceptionalism and bias. These contentions underscore the complexity and sensitivity of engaging with the concept of ancient singularities and emphasize the need for rigorous scholarship and respectful engagement with diverse perspectives. Addressing these criticisms and controversies requires a multidisciplinary approach that encompasses archaeology, history, anthropology, philosophy, and other relevant fields. Engaging in constructive dialogue, critically examining assumptions, and acknowledging the limitations of contemporary frameworks are crucial steps towards navigating the intricate terrain of ancient technological evolution.

Conclusion

The debate surrounding the concept of ancient singularities is both intriguing and complex. Despite the criticisms and controversies, it is clear that the exploration of technological advancements in ancient civilizations has the potential to reshape our understanding of human history and innovation. As we conclude this chapter, it becomes evident that the identification and analysis of ancient singularities require a nuanced approach that accounts for diverse cultural, social, and environmental factors. The study of ancient singularities offers valuable insights into the trajectory of technological development and its implications for ancient societies.

Transitioning to the next chapter, we will delve deeper into the impact of ancient singularities on contemporary technological advancements. By drawing connections between past and present, we aim to elucidate the enduring influence of ancient technologies on modern society. The exploration of comparative analysis with modern singularity theories will offer a holistic view of technological evolution and its relevance across different historical periods. With this transition, we embark on a journey to bridge the gap between ancient wisdom and contemporary innovation, setting the stage for a thought-provoking exploration of the future of ancient technology.

3

The Dawn of Ingenuity

Introduction to Early Human Innovation

In the annals of human history, the dawn of innovation marks a pivotal moment that catapulted our ancestors towards an era of unprecedented progress and advancement. The narrative of transformative change lies in the cognitive revolution, an epoch characterized by the emergence of complex thought and revolutionary breakthroughs in problem-solving. At the heart of this revolution was a profound shift in the cognitive abilities of early humans; a cognitive leap that endowed them with the capacity to think abstractly, imagine the unseen, and engage in novel forms of creative expression. This evolutionary turning point propelled our forebears beyond the confines of mere survival, igniting within them a relentless curiosity and an insatiable drive to innovate. The initial spark that triggered inventive thinking in early humans can be traced back to their burgeoning capacity for symbolic and abstract thought, laying the foundation for cultural, social, and technological evolution. As these early humans sought to understand and make sense of their environment, they began to create mental representations of their surroundings and conceptualize abstract ideas. This newfound ability to conceptualize not only enabled them to communicate and collaborate effectively but also fostered a propensity for experimentation and problem-solving. It was this innate inclination towards innovation that spawned the genesis of tool-making, artistic expression, and symbolic communication, setting the stage for a monumental technological journey that would reshape the course of human history. Furthermore, the dawn of innovative thought was intricately intertwined with the development of language, as verbal communication became the conduit through which knowledge, ideas, and innovations were transmitted across generations. The inception of linguistic complexity and nuanced communication facilitated the exchange of technical insights, enabling early humans to refine existing technologies, devise novel solutions, and cultivate a collective reservoir of knowledge. Thus, the seeds of ingenuity sown during this transformative period germinated into a rich tapestry of innovation, marking the genesis of human creativity and technological prowess. Indeed, the unique cognitive capabilities and innovative spirit of early humans laid the groundwork for a legacy of trailblazing discoveries, signaling the birth of an extraordinary era defined by visionary brilliance and unparalleled progress.

Cognitive Revolution: Emergence of Complex Thought

The cognitive revolution marked a pivotal turning point in human history, representing the emergence of complex thought processes and cognitive abilities that distinguished Homo sapiens from other species. This remarkable transition saw the development of advanced cognitive functions, including symbolic thinking, imagination, and complex problem-solving skills. The roots of this revolution can be traced back to the neurological changes that occurred in the brains of early humans, leading to enhanced capacities for cognitive abstraction and conceptualization. These evolutionary developments laid the groundwork for the unprecedented technological advancements that were to follow.

At the heart of the cognitive revolution was the ability to engage in abstract thinking and mental simulations. This cognitive leap allowed early humans to transcend immediate sensory experiences and envision scenarios beyond their immediate surroundings. Their capacity for imagination opened up new realms of possibilities, enabling them to conceive innovative solutions, devise elaborate plans, and anticipate potential outcomes. This marked a crucial departure from the more instinct-driven behavior observed in preceding hominid species, as it empowered Homo sapiens with the ability to create and implement complex strategies based on mental representations of the world.

The cognitive revolution also facilitated the development of sophisticated social structures and cultural norms. As early humans began to communicate complex ideas and form intricate belief systems, they forged stronger social bonds and collective identities. Language played a central role in this process, serving as a vehicle for sharing knowledge, transmitting cultural practices, and collectively envisioning future endeavors. Through language, humans could articulate abstract concepts, recount historical events, and communicate visions of the future, thereby fostering a shared understanding of the world and stimulating collaborative innovation.

Furthermore, the cognitive revolution precipitated a significant shift in the nature of human cognition, paving the way for an unprecedented level of specialization and division of labor within early societies. With the ability to engage in complex thought processes, individuals could develop specialized skills, pursue diverse areas of expertise, and contribute uniquely to the advancement of their communities. This diversification of cognitive abilities enabled synergistic collaborations, propelling the evolution of increasingly sophisticated technologies, tools, and societal structures.

In essence, the cognitive revolution laid the cognitive foundations for the unparalleled ingenuity and creativity that defined the subsequent phases of technological innovation. By expanding the boundaries of human thought and imaginative capacities, it set the stage for the development of increasingly intricate and transformative technologies, laying the groundwork for the extraordinary progress that was yet to come.

The Role of Language in Technological Development

Language stands as one of the most remarkable and defining hallmarks of human civilization, serving as the bedrock upon which technological progress has been built. Developing a sophisticated means of communication allowed early humans to share knowledge, collaborate on tasks, and pass down accumulated wisdom from generation to generation. This facilitated the refinement of tools, the implementation of innovative techniques, and the organization of complex societal structures. Through language, humans were able to describe their surroundings, articulate abstract concepts, and devise strategies for overcoming challenges. The ability to communicate effectively paved the way for advancements in various fields such as agriculture, architecture, and craftsmanship.

Furthermore, as language evolved, so did humanity's capacity for invention. The development of symbolic communication gave rise to the recording of knowledge, enabling individuals to document their discoveries, achievements, and failures. This laid the foundation for the accumulation of collective expertise, leading to the perpetual refinement and dissemination of technological expertise. Additionally, language enabled the formation of narratives, myths, and oral traditions that often encapsulated valuable insights and lessons about the natural world and its phenomena.

Beyond oral communication, the emergence of written languages further accelerated technological development. Writing systems allowed for the creation of permanent records, facilitating the preservation and transmission of knowledge across vast temporal and spatial distances. This archiving of information permitted the improvement and standardization of techniques, the codification of laws and regulations, and the documentation of empirical observations.

Moreover, the exchange of ideas and the cross-pollination of knowledge through linguistic interactions among diverse cultures catalyzed breakthroughs in technological innovation. As different linguistic communities intersected, they shared unique perspectives, methodologies, and discoveries, leading to the assimilation and adaptation of novel approaches. The amalgamation of various linguistic conventions engendered a rich tapestry of inventive practices and problem-solving strategies.

In conclusion, the role of language in technological development cannot be overstated. From its inception as a tool for verbal expression to its evolution as a medium for preserving and propagating knowledge, language has been indispensable in propelling humankind toward ever more sophisticated forms of technology.

Tool Making: From Simple Stones to Complex Instruments

The evolution of tool making has been a defining factor in the progress of human civilization. As our ancestors ventured beyond basic survival, the development of tools marked a significant leap forward in their ability to manipulate the world around them. Initially, early humans used simple stones as tools, chipping away at rocks to create sharp edges that could be used for cutting, scraping, and hunting. These rudimentary implements laid the groundwork for a gradual transition towards more complex instruments. The mastery of tool making required a deep understanding of raw materials and their properties. It involved experimentation, innovation, and knowledge transfer within communities. Over time, advancements in tool making paralleled the cognitive and social developments of early societies. The transition from simple stones to more elaborate instruments reflected the growing complexity of human interaction with their environment.

Furthermore, the refinement of tools went hand in hand with the diversification of human activities. For instance, the emergence of specialized tools for hunting, farming, and construction paved the way for distinct roles within communities, fostering cooperation and interdependence. With an emphasis on efficiency and precision, the evolution of tools allowed for the exploitation of resources, ultimately impacting settlement patterns and the establishment of permanent dwellings. The utilization of diverse materials, such as bone, wood, and plant fibers, contributed to the creation of increasingly sophisticated tools. This multifaceted approach broadened the scope of human capability, leading to innovations in agriculture, craftsmanship, and creative expression.

As societies progressed, the craftsmanship associated with tool making became a symbol of cultural identity and technological prowess. The transmission of knowledge across generations resulted in the preservation of traditional techniques and the continuous enhancement of tool-making skills. The burgeoning diversity in tools mirrored the richness of human experiences, encompassing a wide array of designs and functions. From hand axes to composite tools, humanity's ingenuity was manifested through the creation of diverse and specialized implements. The interconnectedness of tools, culture, and society underlined the integral role of tool making in the intricate tapestry of human history.

Social Structures and Their Influence on Technology

Social structures play a pivotal role in the development and evolution of technology. The organization of human communities, their division of labor, and the distribution of power within societal frameworks significantly influence the trajectory of technological advancement.

One key aspect to consider is the specialization of labor within early societies. As communities grew and diversified, individuals began to focus on specific tasks, leading to the emergence of professions and trades. This division of labor not only enhanced efficiency but also spurred innovation as individuals sought to improve their respective crafts. The societal recognition of particular skills and expertise also incentivized the refinement and expansion of technological practices.

Moreover, the hierarchical nature of social structures in ancient civilizations often exerted a profound impact on technological progress. Rulers, aristocrats, and other influential figures frequently directed resources towards ambitious technological projects, such as the construction of monumental architecture or the development of advanced weaponry. These endeavors often pushed the boundaries of existing knowledge and capabilities, driving technological innovation through ambitious state-sponsored initiatives.

Additionally, social structures inherently shape the transmission of knowledge and skills across generations. In tightly knit communities, traditional crafting techniques and technological know-how were passed down through oral traditions and apprenticeships, preserving historical methods while allowing for incremental improvements to be made over time. On the other hand, expansive empires facilitated the dissemination of groundbreaking inventions across vast territories, leading to the widespread adoption of new technologies and fostering cross-cultural exchanges that further enriched technological landscapes.

Furthermore, the role of belief systems and cultural practices cannot be overlooked when examining the interaction between social structures and technology. Spiritual beliefs, rituals, and societal values often inspired the development of specialized tools, architectural marvels, and innovative agricultural techniques. The alignment of technology with cultural activities reinforced its significance within the fabric of society and prompted continuous experimentation and refinements based on evolving needs and belief systems.

In conclusion, the interplay between social structures and technology is a multifaceted and dynamic relationship that underpins the progress of human civilization. By recognizing the impact of social organization, division of labor, power dynamics, knowledge transmission, and cultural influences, we can gain a deeper understanding of how ancient societies harnessed their collective ingenuity to drive technological innovation and shape the course of history.

Fire Mastery and Its Impact on Civilization

The mastery of fire marked a pivotal point in human history, shaping the trajectory of civilization and dramatically impacting our technological evolution. The control of fire not only provided warmth, light, and protection from predators, but also facilitated significant advancements in food preparation, leading to improved nutrition and health. From a technological standpoint, the ability to harness and manipulate fire revolutionized tool-making and manufacturing processes. Early humans utilized fire to harden spears, shape tools, and fashion containers for storage and cooking. The development of ceramic materials, such as pottery, was made possible through the controlled application of fire, laying the foundation for complex societies. Beyond these practical applications, the symbolic significance of fire cannot be overstated. Fire became a focal point for communal gatherings, rituals, and spiritual expression, gradually shaping early social structures and belief systems. The visibility of fire at night served as a powerful beacon for distant tribes, enabling communication and interaction between different groups. In this way, the mastery of fire fostered the growth of interconnected communities and the exchange of knowledge and ideas. The control of fire also played a crucial role in the expansion of human habitats. The ability to clear land through controlled burns allowed for the cultivation of agricultural fields and the establishment of settlements, leading to the emergence of sedentary lifestyles. This transition was instrumental in fostering specialized labor, trade, and the development of early forms of governance and societal organization. Furthermore, the impact of fire extended to the preservation of cultural knowledge through early forms of record-keeping. By using charred sticks or rocks to create markings, ancient societies could convey information, stories, and events across generations, laying the groundwork for written language and the accumulation of collective wisdom. The mastery of fire not only transformed the physical environment but also served as a catalyst for intellectual and cultural advancement, ultimately shaping the intricate tapestry of human civilization.

Symbolism and Early Record Keeping

The emergence of symbolic representation and early record-keeping techniques marks a pivotal development in the human story.

As Homo sapiens began to populate diverse landscapes, they developed the capacity for abstract thought and expression, paving the way for the birth of artistic endeavors, cultural symbolism, and rudimentary record-keeping systems. This marked a monumental shift in cognitive abilities as humans sought to represent their experiences and communicate complex ideas through non-verbal means.

One of the earliest evidences of symbolic expression can be traced back tens of thousands of years to the Upper Paleolithic period. Cave paintings, petroglyphs, and intricate carvings found at sites like Lascaux in France, Altamira in Spain, and Chauvet in the Ardeche region of France serve as poignant testaments to the advanced cognitive faculties of our ancient ancestors. These depictions not only showcased their intricate understanding of the natural world, but also hinted at the early stages of storytelling, myth-making, and perhaps even spiritual beliefs.

The transition from mere utilitarian tools to objects with symbolic or ceremonial significance is exemplified in artifacts such as the intricately carved Venus figurines, possibly representative of fertility cults or symbolic of female deities. The use of pigments and dyes for body decoration or cave art further underscores the growing importance of symbolism in early human cultures.

As visionaries of their time, these early humans recognized the need to preserve and transmit information across generations. One remarkable extension of this was the advent of early forms of record-keeping. Whether it was tally marks on bones and stones, notches on sticks, or other mnemonic devices, the ability to record numerical data, astronomical events, or seasonal changes indicated the dawn of proto-writing and a significant leap forward in informational storage.

Moreover, the proliferation of engraved and decorated artifacts reflected an emergent societal structure that valued individual and collective expression. This laid the foundation for the rise of artisans, storytellers, shamans, and archivists—those entrusted with safeguarding the collective memories and cultural heritage of their communities. Such reverence for memory and communication through symbols represents a crucial juncture in human history, setting the stage for the future evolution of written language and complex social structures.

In essence, the incorporation of symbolism and early record-keeping ushered in an era of cognitive expansion and cultural innovation, signifying a remarkable leap forward in human intellectual and expressive capabilities. It laid essential groundwork for the development of languages, belief systems, and organized societies, offering insights into the creative intellect of our ancient forerunners.

Technological Milestones of the Upper Paleolithic

The Upper Paleolithic period, spanning from approximately 50,000 to 10,000 years ago, marked a pivotal phase in human technological development. During this era, our ancestors demonstrated remarkable creativity and adaptability, leading to numerous technological milestones that transformed their way of life. One of the most striking advancements was the innovation of stone tools, evolving from the crude hand axes of the Lower Paleolithic to the finely crafted blades and spearheads of the Upper Paleolithic. These advancements were facilitated by the refinement of knapping techniques, allowing for the production of tools with greater precision and efficiency. The introduction of composite tools, such as spear-throwers and harpoons, showcased an understanding of leverage and projectile mechanics, demonstrating a sophisticated grasp of engineering principles. Additionally, evidence of specialized toolkits for specific tasks suggests a level of foresight and planning previously unseen. Beyond stone tools, the Upper Paleolithic saw the emergence of other innovative technologies. The production of pigments for cave art and body decoration signified a symbolic and aesthetic awareness, while the creation of jewelry from shells and animal bones indicated an early form of self-expression and social identity. Furthermore, the development of bone and antler tools, including needles and awls, revolutionized textile production and enabled more complex forms of clothing and shelter. This era also witnessed the utilization of new materials such as ivory, providing insights into experimentation and adaptation in response to changing environments and resources. The proliferation of these technological advancements during the Upper Paleolithic not only enhanced human survivability but also laid the foundation for future innovations. The intricate designs and precise craftsmanship evident in these artifacts underscore the cognitive and manual dexterity of our ancient predecessors, dispelling any notion of them as primitive or unsophisticated. Instead, their achievements testify to a sophisticated understanding of natural materials and their potential uses, indicative of a burgeoning technological intelligence that would continue to propel human progress.

Transition to Settled Life: Neolithic Innovations

The transition to settled life during the Neolithic period marked a pivotal moment in the course of human history, characterized by a series of remarkable technological and societal innovations. As nomadic hunter-gatherer communities gradually embraced agriculture and animal husbandry, they laid the foundation for permanent settlements and the development of early civilizations. One of the most significant Neolithic innovations was the domestication of plants and animals, which led to the establishment of agricultural practices and the cultivation of staple crops such as wheat, barley, rice, and maize. This shift from a predominantly nomadic lifestyle to sedentary existence brought about fundamental changes in social organization, economic systems, and technological advancements.

In addition to agriculture, the Neolithic period witnessed the emergence of pottery, a groundbreaking innovation that revolutionized food storage, cooking techniques, and the creation of durable tools. The crafting of pottery vessels not only addressed practical needs but also demonstrated advancements in material culture and artistic expression. Moreover, the proliferation of pottery artifacts has provided valuable insights into ancient societies' subsistence patterns, trade networks, and cultural traditions.

The development of architectural techniques also played a pivotal role in the transition to settled life. Early Neolithic communities constructed permanent dwellings using a variety of building materials, including mud bricks, stone, and timber. These architectural innovations facilitated the creation of more complex social structures and communal spaces, laying the groundwork for urban planning and infrastructure development in subsequent civilizations.

The invention of various agricultural tools, such as the plow and irrigation systems, further enhanced productivity and enabled larger-scale food production. This increased agricultural output not only supported burgeoning populations but also fostered specialized labor and the exchange of goods, contributing to the rise of trade and commerce within and across Neolithic societies.

The transition to settled life during the Neolithic era was not solely confined to technological advances; it also ushered in new religious, symbolic, and ritualistic practices, exemplified by the construction of megalithic structures and the evolution of belief systems. These cultural developments reflected the increasing complexity of human societies and the deepening connections between people, their environment, and spiritual beliefs.

In summary, the Neolithic innovations that accompanied the transition to settled life fundamentally transformed human existence, paving the way for the emergence of advanced civilizations and the enduring impact of technology on the trajectory of human progress.

Conclusion: Setting the Stage for Mesopotamian Advances

The Neolithic innovations laid a robust foundation for the subsequent technological advancements witnessed in Mesopotamia. As communities shifted from nomadic lifestyles to settled agricultural practices, their understanding of land cultivation, animal domestication, and irrigation systems burgeoned. These developments marked the genesis of organized societies, setting the stage for a surge in specialized labor, social hierarchy, and a burgeoning need for sophisticated tools and structures. The establishment of permanent settlements fostered the growth of Mesopotamian culture, bestowing upon it crucial prerequisites for engineering feats and scientific progress. Moreover, the compendium of knowledge and experience acquired during the Neolithic era formed the bedrock upon which Mesopotamian scholars and inventors would build and refine their craft.

One cannot overlook the pivotal significance of the inventive spirit cultivated during the Neolithic epoch in preparing the ground for Mesopotamian advances. The emergence of pottery, textile production, and metallurgy not only broadened the spectrum of available materials but also encouraged further experimentation and innovation. The skillful manipulation of minerals and the fabrication of more durable tools granted early artisans the capability to explore increasingly complex designs and utilities. This period of creativity and exploration instilled in individuals a profound appreciation for resource management and problem-solving, two fundamental tenets that became ingrained within the ethos of Mesopotamian society and were upheld by its craftsmen and engineers.

Furthermore, the advent of writing systems and numerical notations among the Neolithic communities elevated the capacity for record-keeping and data preservation. This cultural development, intertwined with economic systems and trade, set the precedent for the sophisticated administrative apparatus that later flourished in Mesopotamia. The meticulous tracking of resources, inventories, and transactions provided fertile ground for the establishment of urban centers with centralized governance, heralding the ascension of the ancient city-states. These socio-economic mechanisms foreshadowed the grandeur of Babylonian and Assyrian empires, where such infrastructures matured into formidable administrative frameworks managing vast territories and sustaining complex economies.

In conclusion, the Neolithic period irrefutably sowed the seeds for the remarkable technological, societal, and intellectual achievements witnessed in Mesopotamia. The blueprint laid out by those early innovators encompassed an amalgamation of knowledge, skills, and cultural practices that collectively propelled human civilization towards greater heights. It is within this transformative context that Mesopotamian advances unfolded, forever indebted to the enduring legacy of humanity's earliest ingenuity.

4

Civilization's Cradle: Technology in Mesopotamia

Mesopotamia: The Birthplace of Urban Tech

Mesopotamia, often referred to as the 'cradle of civilization', is celebrated for being the birthplace of early urban technologies that laid the foundation for modern society. Situated in the fertile crescent between the Tigris and Euphrates rivers, the region's geographical features offered the perfect conditions for agricultural development and the growth of city-states. As one of the earliest known complex societies, the area now encompassing Iraq, Kuwait, parts of Syria, and Turkey was home to a rich and diverse culture that flourished around 3500 BCE. The innovation and resourcefulness of the Mesopotamian people were crucial in paving the way for advancements in agriculture, governance, writing, and various other aspects of civilization. The physical characteristics of the land, with its rivers and abundant natural resources, provided the necessary ingredients for the inception of an advanced society. The Tigris and Euphrates rivers not only facilitated irrigation but also sustained a flourishing ecosystem that supported a burgeoning population. This favorable environment spurred the development of sophisticated farming techniques, including the creation of irrigation systems such as canals and levees, which enabled the cultivation of extensive fields and the subsequent surplus of food. Moreover, the strategic location of Mesopotamia at the crossroads of trade routes fostered cultural exchange, leading to the assimilation of new ideas, practices, and technologies. The confluence of these factors contributed to the emergence of urban centers, where the convergence of diverse skills and knowledge sparked remarkable developments. These pivotal historical achievements underscore the profound impact of Mesopotamia as the cradle of urban technology, laying the groundwork for the remarkable progress that continues to shape human civilization.

Irrigation and Agriculture: Foundation of Civilization

Agriculture and irrigation formed the cornerstone of early Mesopotamian civilization, laying the groundwork for the development of complex societies. The fertile land between the Tigris and Euphrates rivers, known as the Fertile Crescent, provided an ideal environment for agricultural pursuits. Mesopotamian farmers mastered the art of irrigation to harness the waters of these unpredictable rivers, creating intricate canal systems to control flooding and ensure consistent water supply for their crops. The introduction of irrigation techniques such as the shaduf and the sakia revolutionized farming practices, allowing for the expansion of arable land and increasing agricultural productivity. This shift from a nomadic lifestyle to settled agriculture led to the establishment of permanent settlements and the growth of urban centers, fundamentally altering the socio-economic landscape of the region. The surplus food production resulting from advanced agricultural methods enabled specialization of labor and the emergence of merchants, craftsmen, and governing authorities. Moreover, the surplus not only sustained the burgeoning population but also facilitated trade and commerce with neighboring regions, fostering cultural exchange and technological diffusion. Furthermore, the development of sophisticated irrigation and agricultural practices in Mesopotamia served as a blueprint for subsequent civilizations, influencing the evolution of agrarian societies around the world. The legacy of Mesopotamian agricultural innovation endures, shaping the foundation of modern agricultural practices and the way in which civilizations have sustained themselves over millennia.

Cuneiform Script: Writing Systems as Data Storage

Cuneiform script, the earliest known form of writing, played a seminal role in the development and dissemination of knowledge and information in ancient Mesopotamia. The intricate system of wedge-shaped characters impressed into clay tablets served as a means of recording transactions, contracts, administrative documents, religious texts, and literature, thus demonstrating its significance as a key vehicle for storing and transmitting data. The evolution of cuneiform is intrinsically linked with the rise of complex societies, reflecting the intellectual and cultural advancements of the region. As a writing technology, it not only facilitated the administration of complex social structures but also provided a platform for preserving myths, laws, and scientific knowledge.

The versatility of cuneiform allowed for the representation of multiple languages, including Sumerian, Akkadian, and Assyrian, thereby serving as a unifying medium that transcended linguistic barriers within the diverse communities of the ancient Near East. Its adaptability and resilience over several centuries attest to its enduring impact on civilization. Moreover, the educational reliance on cuneiform conveys the value placed on the transference of knowledge across generations, fostering continuity and collective memory.

One of the landmark achievements of cuneiform was the compilation of the Code of Hammurabi, a comprehensive legal codex inscribed in stone that encapsulated Babylonian laws. This monumental feat exemplifies the essential role of writing systems in disseminating and upholding a society's ethical and legal framework. Furthermore, cuneiform permitted the documentation of celestial observations, leading to significant advancements in astronomy and astrology. The rigorous record-keeping enabled by this writing technology laid the groundwork for the codification and preservation of empirical observations and theoretical understandings, thereby serving as a foundation for the scientific inquiries of future civilizations.

The utilization of cuneiform script as a mechanism for archiving and retrieving information underscores its pivotal role as an early form of data storage, a concept resonant with the digital age. Analogous to contemporary databases, cuneiform tablets functioned as repositories for varied forms of knowledge, ensuring its accessibility and preservation for posterity. In essence, the adoption of cuneiform as a writing system empowered Mesopotamian societies by providing a tangible means for the accumulation and transmission of intellectual and cultural wealth, thereby engendering a legacy of knowledge preservation that continues to inspire contemporary information management practices.

Mathematics and Astronomy: The Tools of Timekeeping and Prediction

Mathematics and astronomy played pivotal roles in the technological advancements of ancient Mesopotamia, laying the groundwork for modern scientific inquiry and understanding. The development of sophisticated mathematical concepts, such as the sexagesimal system, allowed Mesopotamian scholars to make precise astronomical observations and predictions. This mathematical system, based on the number 60, facilitated complex calculations and enabled accurate geometric measurements.

Astrology and astronomy were intricately entwined in Mesopotamian culture, and the study of celestial bodies was fundamental to their cosmological beliefs. Astronomical knowledge was utilized to create calendars for agricultural planning, religious festivals, and administrative purposes. Mesopotamian astronomers meticulously documented celestial events, such as lunar eclipses and planetary movements, contributing to the establishment of a comprehensive understanding of the cosmos.

The construction of observatories, known as ziggurats, served as platforms for astronomical observations, emphasizing the importance accorded to celestial phenomena in daily life. These astronomical viewpoints also provided practical insights into timekeeping, navigation, and seasonal patterns, influencing various aspects of society and governance.

Additionally, the Babylonians developed advanced techniques for tracking the motions of celestial bodies, recognizing recurring patterns and formulating predictive models. By integrating mathematics with astronomical observations, they created the foundation for empirical inquiry and the systematic recording of natural phenomena. The emerging understanding of cyclical phenomena and mathematical relationships paved the way for enduring contributions to astronomy and geometry, setting significant precedents for future scientific exploration.

Furthermore, the application of trigonometry in the analysis of angular distances and the measurements of celestial positions reflected the Mesopotamians' remarkable grasp of mathematical principles. Their innovative methodologies in charting the heavens laid the groundwork for later developments in mathematical and astronomical sciences, exerting profound influence on subsequent civilizations throughout history.

The Wheel and Transportation: Mobility and Economic Expansion

The invention of the wheel in Mesopotamia marked a revolutionary leap in ancient technological innovation, introducing a transformative tool that would forever change human civilization. The concept of using round objects to reduce friction and aid movement revolutionized transportation and facilitated the exchange of goods, services, and information across vast distances. This key advancement in mobility engendered economic expansion, facilitating trade and commerce on an unprecedented scale. The wheel's impact was felt not only in Mesopotamia but reverberated throughout the ancient world, shaping the course of history and setting the stage for interconnected global societies.

The adoption of wheeled transportation systems drastically increased the efficiency of moving goods, resulting in enhanced productivity and economic prosperity. Trade routes expanded, linking distant cities and civilizations, creating intricate networks that facilitated cultural exchange, technological diffusion, and the transfer of knowledge. The development of wheeled vehicles such as carts and chariots transformed the movement of people and goods, allowing for the efficient transport of agricultural produce, raw materials, and luxury items, thus contributing to the burgeoning economy of ancient Mesopotamia. The mobility provided by the wheel also spurred urbanization and the growth of city-states, underscoring the crucial role it played in shaping the social and economic fabric of early civilizations.

Moreover, the widespread adoption of wheeled transportation systems significantly influenced the evolution of military tactics and warfare, enabling the rapid deployment of armed forces and providing strategic advantages on the battlefield. The integration of chariots into ancient warfare had profound implications, altering the dynamics of conflict and conquest, and demonstrating the pivotal role of technological advancements in shaping the course of history. Furthermore, the wheel's introduction as a foundational technology laid the groundwork for subsequent innovations, inspiring further developments in engineering, mechanics, and transportation that would continue to redefine the possibilities and capabilities of ancient societies.

In conclusion, the invention and utilization of the wheel represented a watershed moment in the history of ancient technology, revolutionizing the movement of people, goods, and ideas, and fundamentally transforming the economic landscape of Mesopotamia and beyond. Its enduring legacy resonates through the annals of time, serving as a testament to the ingenuity and resourcefulness of our ancient forebears whose innovative spirit laid the groundwork for the remarkable advancements yet to come.

Architectural Innovations: Ziggurats and City Planning

The architectural innovations of ancient Mesopotamia stand as a testament to the ingenuity and vision of the early urban civilizations. At the heart of this achievement lies the development of ziggurats – massive, stepped structures that served as temples, administrative centers, and focal points of cities. The ziggurat at Ur, for example, is an iconic representation of Mesopotamian architecture, with its terraced layers reaching toward the heavens. These towering monuments were constructed using a blend of artistic skill, engineering prowess, and religious symbolism, reflecting the advanced knowledge and cultural significance of the era. City planning in Mesopotamia also bore profound influence on subsequent urban design. The layout of cities such as Uruk and Ur evidenced advanced concepts of zoning, public infrastructure, and defense systems. The intricate network of canals and streets showcased meticulous organization and foresight, setting the stage for the evolution of modern urban landscapes. Moreover, the creation of monumental structures and strategic urban grids reflected the socio-political structures and centralized governance of Mesopotamian societies. The architectural achievements of ancient Mesopotamia not only highlighted the expertise in construction and engineering but also underlined the cultural, religious, and societal foundations that shaped these marvels. By understanding the complex interplay of architecture, religion, governance, and social order, we can unravel the multifaceted tapestry of ancient Mesopotamia, revealing a civilization whose innovations continue to leave an indelible mark on the course of human history.

Metallurgy and Crafting: From Bronze to Iron

Metallurgy and crafting played significant roles in the technological advancement of Mesopotamia, marking the transition from the Bronze Age to the Iron Age. The mastery of metalworking transformed not only the physical landscape of the region but also its cultural and economic dynamics. Bronze metallurgy, with its alloy of copper and tin, had already been well established by this time, contributing to the production of weaponry, tools, and various artifacts. However, the discovery and utilization of iron ore offered a new chapter in the evolution of materials and manufacturing processes.

The emergence of ironworking revolutionized the development of tools and weapons due to the superior strength and durability of iron-based alloys. This led to an expansion of agricultural capabilities, as more efficient iron plows facilitated improved cultivation techniques and increased food production. Concurrently, advancements in smelting methods and the manipulation of iron ores allowed for the creation of a diverse range of items, from household implements to architectural components, further enriching and diversifying the material culture of the era.

Moreover, the proliferation of ironworking stimulated societal progression by fostering specialization and trade. It engendered the establishment of dedicated artisan communities and workshops, where skilled craftsmen honed their abilities in metalwork, leading to the creation of exquisite ornaments, intricate jewelry, and ornate ceremonial objects. As a result, the artistry and craftsmanship of Mesopotamian ironworkers became renowned throughout the ancient world, serving as a testament to their unparalleled skill and ingenuity.

The introduction of iron also catalyzed shifts in military strategies and warfare, prompting substantial advancements in arms and armor that reshaped the nature of conflicts and conquests. Furthermore, the widespread availability of iron tools and infrastructure materials significantly contributed to the expansion and construction of urban centers, such as fortified city walls and monumental architectural edifices, further solidifying the prominence of Mesopotamia in ancient history.

In essence, the mastery of metallurgy and the transition from bronze to iron marked a pivotal epoch in the technological narrative of Mesopotamia, exemplifying the capacity of human innovation to leverage newfound resources and materials for the betterment of civilization. The transformative impact of these advancements reverberated far beyond the confines of Mesopotamia, influencing the course of human history and setting the stage for subsequent technological breakthroughs and social transformations.

Trade Networks and Early Economics: The Proto-Globalization

The ancient Mesopotamian civilization was not only a cradle of technological advancement but also a thriving center of trade networks and early economic systems. Situated between the Tigris and Euphrates rivers, Mesopotamia served as an essential hub for commercial activities, fostering exchange with distant regions and cultures. This flourishing trade was facilitated by the abundance of natural resources in the region, including timber, metals, and agricultural produce. Caravans laden with goods traversed vast distances, connecting Mesopotamia with neighboring lands such as Egypt, the Indus Valley, Anatolia, and the Levant.

One of the key factors driving this proto-globalization was the development of efficient transportation mechanisms, particularly the invention of the wheel. This groundbreaking innovation revolutionized the movement of goods, allowing for larger quantities of merchandise to be transported over greater distances. Furthermore, the emergence of navigable waterways, such as the Euphrates River, significantly enhanced trade routes and expedited commerce across the region. As a result, Mesopotamian merchants were able to engage in extensive transcontinental trade, establishing economic ties with distant civilizations.

The scale of commercial activity in ancient Mesopotamia led to the establishment of complex economic systems. The use of standardized weights and measures streamlined transactions, fostering trust and consistency in business dealings. Additionally, the introduction of cuneiform writing enabled accurate record-keeping and documentation of trade agreements, contributing to the development of rudimentary legal frameworks governing commerce. The efficient administration of trade and financial transactions laid the groundwork for the evolution of early economic principles and practices.

Mesopotamia's burgeoning trade networks not only facilitated the exchange of goods but also facilitated cultural diffusion and the dissemination of knowledge and ideas. Through commercial interactions, the Mesopotamians exchanged technologies, arts, philosophies, and religious beliefs with neighboring civilizations, engendering a rich tapestry of cultural diversity and interconnectedness. This cross-cultural fertilization played a pivotal role in laying the foundation for the globalization processes that would continue to shape the course of human history.

As a testament to its significance, the impact of Mesopotamia's trade networks and early economic systems endures in contemporary society. The enduring legacy of these ancient commercial networks is evident in modern economic practices, global trade relationships, and the interconnectedness of diverse cultures and societies. Understanding the proto-globalization of ancient Mesopotamia provides valuable insights into the historical underpinnings of our present-day globalized world, illustrating the enduring resonance of early economic dynamics and international exchange.

Medicine and Healing Practices: Ancient Health Technologies

Medicine in ancient Mesopotamia was a blend of religious rituals, empirical knowledge, and early scientific methods. The Mesopotamians made significant contributions to the field of medicine, laying the foundation for many practices still used in modern healthcare. One of the most notable aspects of ancient Mesopotamian medicine was the belief in supernatural causes of diseases. They attributed illnesses to malevolent spirits, curses, and divine displeasure, leading to a strong reliance on exorcisms and magical healing rituals. However, alongside these spiritual beliefs, the Mesopotamians also employed practical medical techniques and treatments.

The ancient Mesopotamians were adept at using various natural substances, such as plants and minerals, for medicinal purposes. They developed an extensive materia medica, documenting the properties and uses of numerous herbs and other remedies. Additionally, they practiced surgery and excelled in the art of bandaging and wound management. Physicians of that era were skilled in setting bones, treating infections, and administering different forms of medication. Their expertise in diagnosing and treating diseases, while influenced by superstitions, demonstrated a significant understanding of bodily functions and ailments.

Furthermore, the diagnostic abilities of Mesopotamian physicians were surprisingly advanced for their time. They observed symptoms, made connections between them, and documented their findings meticulously. The famous Diagnostic Handbook, known as the Enuma Anu Enlil, contained detailed descriptions of diseases and their clinical features. This compendium provided valuable insights into the prognosis and treatment of various ailments. Moreover, the ancient Mesopotamians recognized the importance of maintaining public health and sanitation to prevent the spread of diseases. They developed regulations for city planning and waste disposal, showcasing a rudimentary understanding of epidemiology.

The legacy of ancient Mesopotamian medicine extends beyond its medical knowledge and practices. Many of the treatments and therapies they pioneered have influenced the development of medical systems across different cultures and eras. These enduring influences include the concepts of patient care, pharmacopoeia, surgical techniques, and the documentation of medical knowledge. By exploring the health technologies of ancient Mesopotamia, we can gain a deeper appreciation for the origins of modern medicine and the enduring impact of early innovation on the advancement of healthcare.

Legacy and Influence on Modern Technology: Bridging Millennia

The legacy of Mesopotamian technology continues to reverberate through the annals of history, casting its influence far into the realms of modern-day technology. The innovative strides made in medicine and healing practices during the Mesopotamian era have left an indelible mark on the contemporary medical landscape, reshaping our understanding of health, treatment modalities, and the innovative technologies used in modern healthcare. The ancient Mesopotamians' reverence for the healing arts has been a guiding light for the evolution of medical practices and contributed significantly to the development of advanced medical technologies that we rely on today.

The sophisticated methods and techniques employed by the Mesopotamians to treat ailments and injuries paved the way for modern medical innovations and forged the foundation upon which our current healthcare systems are built. The conceptualization of disease as having natural causes, the formulation of herbal remedies, and the establishment of early diagnostic techniques all served as pivotal contributions to the burgeoning field of medical technology. Furthermore, the creation of intricate surgical tools, such as scalpels and probes, showcased the Mesopotamians' adeptness in crafting instruments that enhanced precision and efficacy in medical procedures.

The enduring impact of Mesopotamian medicine is conspicuous across various domains of modern healthcare, ranging from pharmaceutical advancements and surgical instrumentation to diagnostic imaging and therapeutic interventions. Today's pharmaceutical industry owes much to the ancient knowledge of medicinal plants and their curative properties, a testament to how Mesopotamian wisdom has transcended time to shape contemporary drug development and pharmacology. Additionally, the meticulous craftsmanship of surgical implements by the Mesopotamians laid the groundwork for the refinement of surgical tools and equipment, catalyzing the metamorphosis of surgical technology into state-of-the-art apparatus utilized in operating theaters worldwide. Moreover, the prescient insights into anatomy and physiology fostered by Mesopotamian scholars continue to resonate in the modern practice of medicine, guiding medical professionals in their pursuit of understanding the human body and formulating effective treatments.

The integration of Mesopotamian medical principles with cutting-edge technologies underscores the extraordinary interplay between ancient ingenuity and contemporary innovation. By embracing time-honored healing philosophies and adapting them to modern methodologies, medical practitioners perpetuate the enduring legacy of Mesopotamian health technologies, paying homage to the formidable expertise of their ancient predecessors. Indeed, the bridge connecting millennia-old medical expertise with present-day technological breakthroughs stands as a testament to the enduring relevance of Mesopotamian innovation, illuminating a path of continuous progression and enlightenment in the sphere of medical technology.

5

Pyramids and Processing: AI in Ancient Egypt

Egyptian Innovations

Ancient Egypt stands as an enduring testament to human ingenuity, where the extraordinary achievements of technology and science continue to captivate and inspire. From the remarkable precision of their architectural feats to the sophisticated understanding of celestial mechanics, the Egyptians displayed a profound mastery of knowledge and skill that reverberates through the annals of history. At the heart of their innovation lay a profound connection between their technological developments and the cultural, spiritual, and scientific ethos of their civilization. The grandeur of the pyramids, the intricate design of temples, and the mystical alignment to astronomical phenomena all bear witness to the profound interconnectedness of Egyptian innovation and their quest for understanding the cosmos.

Moreover, the Egyptian civilization's mastery of engineering and construction techniques represented a true paradigm shift in ancient world technology, unveiling a level of geometric precision that surpassed contemporaneous civilizations. The utilization of advanced tools and methods to achieve perfect alignments with celestial bodies further showcases the profound acumen of the ancient Egyptians in uniting practical knowledge with metaphysical concepts. Understanding these advancements not only illuminates the technical prowess of this ancient civilization, but also provides a window into their deeply held beliefs and reverence for the natural world.

Furthermore, the development and mastery of hieroglyphic scripts and administrative record-keeping reflect an early form of information processing and storage, providing crucial insights into the early seeds of what would eventually be recognized as artificial intelligence. The systematic organizational structure of their administrative offices and the encoding of knowledge within hieroglyphic inscriptions demonstrated their foresight in creating and managing vast repositories of information, laying foundations for future endeavors in data management and information retrieval.

Traversing the intricacies of Egyptian innovation leads us to acknowledge that their prodigious advances serve as an unparalleled source of inspiration and wisdom that transcends the constraints of time. As we delve deeper into the annals of Egyptian history and unravel the mysteries of their innovative spirit, we uncover a tapestry woven with the threads of grandeur, intellect, and divine reverence for the cosmos.

Geometric Precision: Alignment to the Stars

The ancient Egyptians were profound astronomers and resourceful mathematicians, their understanding of celestial movements and geometry was astonishing for their time. The alignment of the pyramids with the stars is a testament to their advanced knowledge and meticulous calculations. Through careful observation of the heavens, the Egyptians were able to construct these monumental structures with incredible precision. The orientation and layout of the pyramids at Giza are thought to reflect specific constellations, serving as not only architectural marvels but also celestial observatories. Scholars believe that the angles and dimensions of the pyramids were purposefully designed to align with certain stars and stellar patterns, showcasing the Egyptians' reverence for cosmic order and their quest for eternal harmony. The intricate relationship between the pyramids and the stars not only highlights the technical prowess of the ancient Egyptians but also underscores their spiritual connection to the cosmos. Moreover, the alignment to the stars suggests a sophisticated understanding of the Earth's position in the universe and the cyclical nature of time. By studying the cosmic movements and incorporating them into their architectural endeavors, the ancient Egyptians demonstrated an early form of interdisciplinary knowledge, fusing astronomy, mathematics, and engineering to produce enduring monuments that still captivate and inspire awe today. The precise alignment also raises questions about the specific techniques employed, igniting scholarly debates and investigations to uncover the exact methods and instruments used in achieving such astronomical accuracy. This intersection of science, spirituality, and craftsmanship continues to intrigue modern researchers and enthusiasts alike, offering a compelling window into the ingenuity and ingenuity of our ancient predecessors.

Engineering the Pyramids: Tools and Techniques

The construction of the Egyptian pyramids stands as a testament to ancient engineering prowess, showcasing astonishing precision and innovative techniques. The immense scale and geometric complexity of these structures required an advanced understanding of mathematics, physics, and materials science. Ancient Egyptian architects and builders utilized a range of tools and methods that demonstrated remarkable sophistication for their time.

One of the critical aspects of pyramid construction was the transportation and manipulation of massive stone blocks. The Egyptians employed skilled laborers, as well as specially designed sledges and ramps, to move and position these colossal stones with incredible accuracy. This process not only required brute force but also a deep comprehension of leverage, friction, and the distribution of weight.

Additionally, the use of water and lubricants likely played a crucial role in facilitating the movement of these heavy objects.

Furthermore, the precision in aligning the pyramids with cardinal directions and celestial bodies suggests a deep understanding of astronomy and geometry. The architects' ability to navigate the landscape and determine the exact positioning for each component of the pyramid demonstrates a level of spatial awareness that is truly impressive. The alignment with astronomical phenomena underscores the Egyptians' reverence for the heavens and their desire to integrate cosmic principles into their architectural plans.

In terms of construction techniques, the Egyptians developed sophisticated methods for quarrying, shaping, and assembling the enormous limestone and granite blocks. Using copper and bronze tools, they were able to cut, carve, and polish the stones with remarkable finesse. The crafting of precise angles and surfaces required exceptional craftsmanship and a deep knowledge of material properties. These techniques allowed the builders to achieve the smooth, flat surfaces and tight-fitting joints that characterize the pyramid's facades.

The construction of the Egyptian pyramids was a multifaceted endeavor that combined technical skill, theoretical knowledge, and organizational acumen. The integration of advanced tools, mathematical principles, and innovative methods exemplifies the ingenuity of the ancient Egyptian civilization. This rich tradition of engineering has left an indelible mark on the landscape and continues to captivate and inspire generations of scholars, engineers, and enthusiasts around the world.

Artificial Intelligence Concepts in Hieroglyphics

The ancient Egyptians employed a complex system of writing known as hieroglyphics, which served not only as a means of communication but also as a medium for preserving knowledge and religious beliefs. Upon closer examination, it becomes evident that some hieroglyphic symbols contain elements that hint at the potential for abstract thought and rudimentary artificial intelligence concepts. Through a multidisciplinary approach involving linguistics, archaeology, and computing, scholars have endeavored to unravel the possibility of AI concepts embedded within the hieroglyphic script. One intriguing aspect is the presence of recurrent patterns and clusters of symbols, reminiscent of algorithmic structures used in modern programming languages. It is conceivable that the arrangement and repetition of certain hieroglyphs may have represented basic computational logic or decision-making processes. Furthermore, the use of hieroglyphic determinatives—symbols that accompany and clarify the meaning of other glyphs—suggests an early form of symbolic representation akin to the encoding and decoding of data in contemporary AI systems. The presence of hieroglyphic codes associated with administrative functions and mathematical concepts also provides compelling evidence of the ancient Egyptians' grasp of abstract and logical reasoning, echoing the principles underpinning artificial intelligence. Moreover, the role of scribes in crafting and interpreting hieroglyphic texts parallels the functions performed by AI programmers and analysts in today's technological landscape. These professionals were tasked with understanding complex algorithms and transforming them into practical applications, demonstrating a remarkable parallel between the ancient scribe's cognitive processes and the methodologies employed in artificial intelligence development. Indeed, the study of hieroglyphics opens a window into the cognitive capabilities and intellectual achievements of ancient Egyptians, shedding light on their potential engagement with proto-AI concepts. This inquiry not only deepens our understanding of ancient civilizations but also prompts reflection on the continuity of human ingenuity in the pursuit of artificial intelligence.

Administrative Automation of the Pharaohs

The administrative systems of ancient Egypt were remarkably advanced for their time, showcasing a level of procedural sophistication that paralleled modern organizational structures. At the heart of this administrative prowess were the pharaohs, who implemented sophisticated bureaucratic mechanisms to govern their vast empire and manage complex societal functions. The unique amalgamation of hieroglyphic writing, numerical records, and advanced organizational structures laid the foundation for what could be considered an early form of administrative automation. In the process of governing their lands, the pharaohs utilized a highly organized system of record-keeping, resource management, and task delegation. This involved the careful documentation of agricultural production, strategic storage of food reserves, and meticulous allocation of resources to sustain the population amid changing environmental conditions. It is fascinating to note how the intricate nature of these administrative tasks required systematic solutions, which likely involved the utilization of early algorithmic thinking. The implementation of these administrative processes allowed for efficient decision-making within the Egyptian bureaucracy, ensuring stability across diverse regions and enabling the kingdom to thrive in an often unpredictable environment. Additionally, the deployment of specialized scribes and overseers further contributed to the regulation and oversight of these administrative systems, creating a hierarchical structure that managed information flow and regulated societal functions. Examining this ancient example of administrative automation offers intriguing insights into the ways in which early civilizations harnessed the power of structured data, precise organization, and strategic planning to maintain societal equilibrium. Furthermore, such historical precedents serve as a testament to humanity's enduring quest for practical and innovative problem-solving, demonstrating our persistent commitment to improving the efficiency of complex operations through controlled and automated processes.

Legacy Data Structures from the Nile

Ancient Egypt boasts a rich legacy of data structures that reflect the civilization's advanced understanding of organization and information management. The Nile served as the lifeblood of Egyptian society, and this profound connection between water and life influenced their approach to recording and preserving knowledge. One of the most enduring and remarkable data structures from ancient Egypt is the hieroglyphic script, a complex system of writing that encompassed both logographic and alphabetic elements. This sophisticated writing system was used for monumental inscriptions, administrative documents, religious texts, and more, demonstrating the diverse applications of their data storage methods. The meticulous preservation of information on papyrus scrolls and temple walls exemplifies the ancient Egyptians' commitment to maintaining an enduring record of their culture and achievements. Beyond textual records, the intricate notations within architectural structures and sacred sites also functioned as unique data structures, encoding symbolic representations of cosmological beliefs, mythology, and divine concepts. These intricate data structures were embedded into the very fabric of Egyptian architecture, serving as repositories of cultural knowledge and spiritual wisdom. Through these enduring manifestations of ancient data structures, we gain valuable insights into the sophistication and intellectual prowess of the ancient Egyptian civilization. By studying and deciphering these legacies, modern scholars continue to unravel the complexities of ancient Egyptian society, shedding light on their innovative approaches to data management, communication, and philosophical thought. These enduring data structures from the Nile region offer a compelling testament to the lasting impact of ancient civilizations on the evolution of human knowledge and technology.

Cognitive Symbolism in Sacred Architecture

The architectural wonders of ancient Egypt offer profound insights into the cognitive symbolism embedded within sacred structures. From the majestic temples to the enigmatic pyramids, the design and layout reflect a deep understanding of the human psyche and spiritual connection. The meticulous alignment of these structures with celestial bodies was not merely a demonstration of astronomical knowledge; it also served as a means of integrating cosmic understanding with the human experience. The deliberate incorporation of symbols and hieroglyphs throughout the architecture further emphasizes the cognitive purpose of these sacred spaces. Each hieroglyphic inscription and symbolic motif was strategically positioned to evoke specific emotions and facilitate transcendental experiences. This intentional placement of symbols aimed to guide individuals through a transformative journey, merging the physical realm with the realm of divine consciousness. Moreover, the intricate patterns and geometric precision employed in construction carried significant psychological implications. The harmonious proportions and symmetrical forms were intended to induce a sense of tranquility and spiritual balance within those who interacted with these architectural marvels. Additionally, the strategic use of light and shadow within these spaces was not just for practical illumination, but also to create an atmosphere that evoked contemplation and reverence. Beyond their structural functionality, the Egyptian sacred architectures acted as elaborate tools for shaping the consciousness of those within their embrace. Their elaborate nature reflects an advanced understanding of the human mind and its potential for spiritual awakening. As we delve deeper into the cognitive symbolism present in these architectural wonders, we uncover a profound testament to the sophistication of ancient Egyptian civilization's insight into the human psyche and its connection to the cosmic order.

Connecting Cosmic Understanding with Computation

Ancient Egyptian civilization was deeply connected to cosmic phenomena, and this profound understanding permeated various aspects of their culture, including their approach to computation. The Egyptians closely observed the movements of celestial bodies and incorporated this knowledge into their architectural designs, religious sites, and everyday life. This cosmic awareness was not merely philosophical or spiritual but also had practical implications in their technological advancements.

The alignment of the pyramids with astronomical features highlights the integration of cosmic understanding with computation. The precise orientation of the pyramids with respect to cardinal directions and astronomical alignments suggests a sophisticated grasp of celestial mechanics. It is evident that the ancient Egyptians possessed advanced knowledge of astronomy and utilized this information to harmonize their monumental structures with the celestial realm.

Furthermore, the intricate calendar systems developed by the Egyptians reflected their deep connection with cosmic cycles. The ability to track and predict celestial events such as solstices, equinoxes, and lunar phases showcased their mathematical prowess and computational acumen. These calendrical calculations not only facilitated agricultural activities but also underscored the intersection of cosmic understanding and computational proficiency.

In parallel with their cosmic insights, the Egyptians employed numerical symbolism and hieroglyphic representations to encode astronomical and cosmological concepts. The use of symbolic imagery for celestial bodies, deities, and cosmic forces conveyed a profound understanding of cosmic interconnectedness. Moreover, these symbolic depictions encapsulated computational concepts within the framework of religious and cultural expressions, demonstrating the fusion of cosmic understanding with artistic and communicative forms.

The integration of cosmic understanding with computation offers valuable insights for contemporary technological endeavors. By recognizing the ancient Egyptian paradigm, modern researchers and innovators can appreciate the holistic approach to technology that transcends mere functionality and embodies a deeper cosmic connection. This holistic perspective not only enriches the development of computational systems but also nurtures a broader societal consciousness rooted in cosmic harmony and technological advancement.

Insights into Egyptian Cryptology

The study of ancient Egyptian cryptology offers a captivating glimpse into the intricate methods used by this advanced civilization to safeguard their knowledge and records. The Egyptians were masterful at encoding information for various purposes, from protecting royal secrets to preserving sacred texts. Their cryptographic techniques reflect a sophisticated understanding of symbols, ciphers, and steganography. Hieroglyphics, the iconic script of ancient Egypt, not only served as a form of communication but also concealed layers of meaning through its complex structures and symbolism. From religious rituals to administrative records, cryptology played a pivotal role in concealing sensitive information from adversaries or unauthorized individuals. The Egyptian understanding of encryption was intrinsic to their societal and mystical practices, intertwining the esoteric with the pragmatic. Deciphering Egyptian cryptology requires an interdisciplinary approach, drawing from linguistics, archaeology, mathematics, and cultural studies. The interplay of hieroglyphic translations, numerical systems, and historical context unveils the nuanced nature of Egyptian encryption. Understanding the evolution of cryptological practices in ancient Egypt provides invaluable insights into the cognitive development of early civilizations and their approach toward securing knowledge. Furthermore, unlocking the cryptographic methodologies of ancient Egypt can potentially yield parallels to contemporary cryptography, offering fresh perspectives on modern data security and encryption algorithms. The enigmatic nature of Egyptian cryptology continues to intrigue scholars and enthusiasts, as unraveling these ancient codes holds the promise of uncovering lost wisdom and hidden narratives from one of the world's most enigmatic cultures.

Summary and Implications for Modern AI

The study of ancient Egyptian cryptology provides intriguing insights into the sophistication of their cognitive abilities and problem-solving techniques. By examining the intricate hieroglyphic scripts, we gain an understanding of how the ancient Egyptians drafted complex writings, potentially depicting advanced symbolic representations beyond mere communication. This raises questions about the potential application of ancient encoding methods as inspirations for modern AI algorithms.

The ability of the Egyptians to encode and protect information through cryptology demonstrates a level of intelligence that aligns with the foundational principles of artificial intelligence. Their strategic use of symbolism and encrypted messaging suggests an early form of data security and transmission standards. The ancient practices, when viewed through a modern lens, offer a glimpse into the cognitive evolution of human thought processes – a crucial consideration for developers and researchers seeking innovative pathways for AI development.

Moreover, the utilization of cognition-driven architecture in the construction and orientation of the pyramids reflects a deep understanding of spatial reasoning, celestial mechanics, and computational precision. These technical achievements parallel the fundamental objectives of contemporary AI research, particularly in the realms of computer vision, spatial awareness, and optimization algorithms. By delving into the enigmatic knowledge systems of the ancient Egyptians, researchers can draw parallels between historical wisdom and cutting-edge AI technologies.

Furthermore, the cognitive symbolism embedded within the architectural design of ancient monuments prompts speculation on the potential integration of symbolic representation in modern AI frameworks. The exploration of how the ancient Egyptians incorporated intelligent design principles and abstract concepts into their structural creations may inspire the development of AI algorithms that encompass higher-level cognition and pattern recognition capabilities.

In conclusion, the revelations from studying Egyptian cryptology and the cognitive applications manifested in the construction and symbolism of ancient monuments yield profound implications for modern AI. They offer valuable insights into potential avenues for enhancing AI systems, infusing them with historical wisdom and cognitive paradigms. By fusing ancient ingenuity with contemporary technological advancements, we have the opportunity to reshape the trajectory of AI research and development, unlocking new frontiers in cognitive computing and symbolic AI applications.

6

Eastern Enigmas: Advanced Tech in Ancient China

Disclosing the Secrets of Ancient Times

As we navigate the annals of time, the sages of ancient China continue to impart their profound wisdom through the transcendent legacies of their technological achievements. The allure of understanding ancient Chinese technology resonates with an ineffable significance in our contemporary era, compelling us to unearth the intricacies of their astronomical and engineering prowess. Through the vantage point of history, tracing the celestial pursuits of early Chinese astronomers unravels a narrative of both empirical observation and philosophical contemplation. The intricate interplay between the heavens and Earth mirrors the enduring human quest for understanding the cosmos, a curiosity that reverberates in the modern pursuit of space exploration and technological innovation. By delving into the foundations of ancient Chinese astronomy, we discern the meticulous calibration of instruments and the sublime precision exhibited in their conceptualization of celestial mechanics. This exploration bridges the chasm of time, providing illuminating insights into the origins of scientific inquiry and technological ingenuity. Moreover, the indomitable mastery of engineering marvels in ancient China surpasses mere physical constructs, birthing a continuum of innovation that echoes across the corridors of time to inspire modern advancements. From the ethereal elegance of the Silk Road to the robust metallurgical conquests encapsulated in bronze artifacts, each facet of ancient Chinese technology serves as a testament to the symbiotic relationship between human intellect and empiricism. Amidst the verdant landscapes of antiquity lie reservoirs of knowledge waiting to be discovered, dissected, and embraced as reservoirs of inspiration. It is within this context that the revival of ancient Chinese technology assumes its mantle of profound relevance in our contemporary world, propelling us to glean from the collective wisdom of bygone eras and to enrich our present endeavors with the gleanings of antiquity.

The Harmony of Heavens: Astronomy and Engineering

Ancient China's profound connection with the cosmos has been a driving force behind its advancements in astronomy and engineering. The celestial bodies held great significance for the early Chinese civilizations, influencing not only spiritual and cultural practices but also technological developments that shaped the course of history. Embracing the concept of harmony with nature, ancient Chinese astronomers meticulously observed the movements of the stars, planets, and constellations, leading to the creation of precise astronomical instruments such as the armillary sphere and celestial globe. These innovations allowed them to develop complex calendars, predict celestial events, and navigate vast seas with remarkable accuracy, laying the foundation for future scientific endeavors.

Moreover, the intricate knowledge of celestial bodies spurred groundbreaking engineering achievements. The construction of awe-inspiring architectural marvels, such as the astronomical observatory at Gaocheng and the renowned equatorial sundial in Dengfeng, showcased the fusion of astronomy and engineering expertise. These structures not only served as practical tools for observing celestial phenomena but also symbolized the integration of celestial harmony into the fabric of everyday life in ancient China.

Furthermore, the relationship between astronomy and engineering extended to agricultural practices and urban planning. Utilizing astronomical knowledge, the ancient Chinese developed sophisticated irrigation systems based on lunar cycles and seasonal changes, maximizing crop yields and sustaining their burgeoning population. Additionally, the layout of cities and towns was often intricately aligned with celestial principles, epitomizing the profound influence of astronomy on the physical environment.

The enduring legacy of Chinese astronomy and engineering reverberates through the ages, inspiring countless scientific inquiries and technological innovations. Even today, the spirit of discovery and reverence for the cosmos continues to shape our understanding of the universe and drive progress in fields as diverse as astrophysics, architecture, and urban design. By delving into the harmonious relationship between ancient Chinese astronomy and engineering, we gain valuable insights into the interconnectedness of human ingenuity, cultural heritage, and the celestial realm, fostering a deeper appreciation for the timeless pursuit of knowledge and harmony.

Silk and Beyond: Innovations in Textile Manufacturing

Textile manufacturing in ancient China is a testament to the sophistication and ingenuity of this remarkable civilization. At the heart of these advancements lies the legendary production of silk, a material so exquisite that it captivated the imagination of the world. However, the textile innovations of ancient China extend far beyond silk alone. The mastery of various materials, dyeing techniques, weaving processes, and garment designs distinguished the ancient Chinese as pioneers in the textile industry.

The production of silk, an art form and science in itself, was shrouded in secrecy, with the methods closely guarded by the Chinese for centuries. The intricate process of cultivating and harvesting silkworms, extracting their precious silk threads, and weaving the fabric required unparalleled skill and dedication. The significance of silk extended beyond its luxurious appeal; it served as a symbol of prestige, wealth, and cultural identity for the ancient Chinese society.

Furthermore, the ancient Chinese explored and excelled in various other aspects of textile manufacturing. They developed advanced techniques for dyeing fabrics using natural pigments derived from plants and minerals, resulting in a vibrant array of colors and patterns. Additionally, the refinement of weaving looms and tools allowed for the creation of intricately designed textiles, incorporating exquisite motifs and detailed craftsmanship.

Beyond the realms of clothing, textiles played a pivotal role in other facets of ancient Chinese culture. Sailcloth, for instance, facilitated maritime exploration and trade, contributing to the expansion of Chinese influence across distant lands. Furthermore, the utilization of textiles extended to furnishing, with elegant tapestries and decorative fabrics adorning palaces, temples, and residences.

The profound impact of ancient Chinese textile innovations reverberates through the annals of history, influencing global trade, cultural exchange, and artistic expression. Their mastery of textile manufacturing not only propelled them to economic prosperity but also enhanced their standing as a beacon of creativity and resourcefulness. Today, the legacy of ancient Chinese textile manufacturing endures, inspiring contemporary artisans, fashion designers, and scholars to continue unraveling its timeless allure and technological achievements. From silk to sailcloth, the ancient Chinese's advancements in textile manufacturing transcend mere fabric production, embodying a rich tapestry of innovation, artistry, and cultural significance.

Reflections of Bronze: Metallurgy and its Mysteries

Bronze, a metal alloy consisting primarily of copper, with significant additions of tin and other elements such as arsenic, has redefined the course of ancient Chinese civilization through its metallurgical prowess. The mastery of bronze casting and forging in ancient China not only revolutionized warfare but also influenced artistic, ceremonial, and ritualistic practices. From the enigmatic bronzes ceremoniously buried in tombs to the finely crafted weapons wielded by military strategists, the art and science of bronze metallurgy unveil a world of fascinating mysteries. The intricate knowledge possessed by ancient Chinese metallurgists is a testimony to their exceptional understanding of materials and their utilization. Through the precise blending and smelting of copper and tin, they were able to create stunning artifacts that represented their cultural, social, and technological advancements. Moreover, the complex techniques used for casting, refining, and decorating bronze objects demonstrate the meticulous skills and innovative approaches adopted by these early engineers. The discovery of elaborate bronze vessels, ornate weapons, and ceremonial tools speaks volumes about the sophisticated craftsmanship and aesthetic sensibilities of ancient Chinese artisans. Furthermore, the utilization of bronze in various ritualistic and religious contexts reflects the spiritual significance attributed to this exceptional alloy. The symbolic connotations associated with bronze objects within the realms of ancestor worship, divination, and ceremonial offerings underscore its profound cultural and societal implications. The inherent mysteries surrounding the production and purpose of these bewitching bronze artifacts continue to captivate modern researchers, offering glimpses into the technological virtuosity and societal complexities of ancient China. Delving deeper into the depths of prehistoric metallurgy, we unravel the secrets of an era shaped by the transformative power of an alloy that transcended mere material significance, enriching the cultural tapestry of a civilization flourishing amidst the enigmatic allure of bronze.

The Art of War: Military Ingenuity

The ancient Chinese civilization displayed an unparalleled level of military ingenuity that continues to fascinate historians and strategists to this day. The evolution of warfare in ancient China was marked by remarkable advancements in weaponry, tactics, and strategic thinking.

At the core of Chinese military prowess lay the development and mastery of various weapons and tools. The invention of gunpowder, for instance, revolutionized the landscape of warfare, introducing the world to the concept of explosive ordnance and propelling projectiles. This significant innovation shifted the balance of power on the battlefield and influenced the course of global conflict for centuries to come. Additionally, the refinement of traditional arms such as swords, spears, and bows exemplified the meticulous craftsmanship and dedication to martial perfection inherent in ancient Chinese culture.

However, beyond the sheer sophistication of their armaments, it was the strategic mindset of Chinese military leaders that truly set them apart. Sun Tzu's 'The Art of War' stands as an enduring testament to the depth of understanding and tactical acumen possessed by ancient Chinese generals. Through profound insights into the psychology of warfare, logistical planning, and the art of deception, this timeless treatise remains a cornerstone of military philosophy, guiding countless strategists and commanders throughout history.

Moreover, the construction of formidable defensive structures, exemplified by the Great Wall of China, illustrates the long-term vision and engineering prowess of ancient Chinese military architects. These monumental fortifications stand as a testament to the unwavering dedication to national defense and the ability to execute grand-scale projects that characterized the ancient Chinese civilization.

The legacy of Chinese military ingenuity extends far beyond the realm of warfare itself. It encapsulates a holistic approach to security, integrating technology, strategy, and governance to safeguard the stability and prosperity of the empire. Furthermore, the principles and tactics developed during this era continue to inform contemporary military thinking, providing invaluable lessons for modern-day strategic planners and military leaders across the globe.

Beneath the Surface: Hydraulic Engineering

Hydraulic engineering in ancient China represented a pinnacle of technological achievement, demonstrating an unparalleled understanding of water management and utilization. The intricate network of canals, levees, and irrigation systems laid the foundation for sustained agricultural development, supporting the burgeoning population and stimulating economic growth. The concept of hydraulic engineering extended beyond mere functionality, mirroring the philosophical tenets of harmony and balance pervasive in Chinese culture. Ingenious devices like the waterwheel were employed to harness natural forces, powering machinery and enhancing productivity. Furthermore, innovative methods of flood control and water diversion showcased a deep comprehension of hydrology and environmental stewardship. These early advancements in hydraulic engineering not only elevated agricultural practices but also contributed to the overall resilience and prosperity of ancient Chinese society. Contemporary scholars continue to be astounded by the sophistication and foresight displayed in these age-old water management systems, recognizing them as a testament to the enduring legacy of ancient Chinese ingenuity.

Seeds of Civilization: Agricultural Advances

Ancient China's agricultural advances laid the foundation for its civilization's prosperity and longevity. The development of sophisticated irrigation systems, including canals, dikes, and water conservation techniques, enabled the cultivation of vast swathes of land, transforming once marginal areas into fertile fields capable of sustaining burgeoning populations. The introduction of innovative farming implements such as the iron plow and harrow revolutionized agricultural practices, making it possible to till and cultivate larger tracts of land with greater efficiency. Furthermore, the ingenious use of crop rotation and early forms of organic fertilization demonstrated an advanced understanding of soil management and sustainability that contributed to bountiful harvests. The meticulous observation and recording of seasonal patterns and weather phenomena facilitated the development of a comprehensive agricultural calendar, enabling precise planning and management of planting, growing, and harvesting activities. Ancient Chinese farmers also engaged in the cultivation of a diverse array of crops, including rice, millet, wheat, and soybeans, showcasing an impressive understanding of crop diversity and its impact on food security and nutrition. Furthermore, the pioneering development of sericulture, or silk production, revolutionized the textile industry and became a significant economic driver for the ancient Chinese civilization. This intricate interplay of technological innovation, ecological stewardship, and agricultural expertise elevated Ancient China's agricultural practices to a remarkably advanced level, contributing to its status as an enduring and prosperous civilization.

Eternal Mountains: Achievements in Architecture

Ancient China stands as a testament to the ingenuity and vision of its architects. The architectural achievements of this era are truly awe-inspiring, reflecting a harmonious blend of functionality, aesthetic beauty, and deep-rooted cultural symbolism. From the majestic Great Wall that spans over thousands of miles to the timeless allure of the Forbidden City, the ancient Chinese created architectural marvels that continue to capture the world's imagination.

At the core of ancient Chinese architecture lies a profound understanding of natural harmony and balance, where structures were meticulously designed to harmonize with the surrounding landscape. This philosophy, known as Feng Shui, emphasized the integration of buildings with nature, promoting the flow of positive energy and creating serene environments. The meticulous attention to detail is evident in the intricate wooden frameworks, grand courtyard complexes, and ornate pagodas that grace the skyline.

Moreover, the ancient Chinese architects displayed remarkable technical prowess in material engineering, utilizing advanced building techniques and innovative materials. Their mastery of timber construction, evident in the exquisite curved roofs and intricate joinery, showcased an unparalleled level of craftsmanship and structural integrity. Additionally, the development of advanced masonry and brickwork allowed for the creation of enduring monumental structures, ensuring the preservation of these architectural wonders for centuries to come.

One cannot overlook the profound spiritual and cultural significance embedded within these architectural masterpieces. The design of temples, palaces, and imperial tombs served not only as functional spaces but also as embodiments of philosophical and religious beliefs. Each element, from the layout and orientation to the intricate ornamentation, encapsulated profound metaphysical principles and dynastic symbolism, reflecting the unity of heaven and earth as well as the enduring spirit of the civilization.

The lasting legacy of ancient Chinese architecture extends beyond its physical manifestations, permeating into modern design, urban planning, and sustainable construction practices. The timeless principles of balance, symmetry, and environmental coexistence remain influential, serving as a source of inspiration for architects and city planners worldwide. As we gaze upon these eternal mountains of stone and wood, we bear witness to the enduring legacy of a civilization whose architectural achievements transcend time, leaving an indelible imprint on the world's architectural landscape.

Digital Dragons: Conceptualizing Computations

Ancient China's advancements in the realm of computation and digital innovation have long fascinated historians, engineers, and visionaries alike. Their intricate understanding of numerical systems and their practical application set the foundation for later developments in mathematics and computer science.

At the heart of ancient Chinese computations was the abacus, a manual device that formed the basis of many complex calculations. Its mastery required an intimate understanding of mathematical concepts and served as a testament to the sophisticated computational abilities of ancient scholars. Moreover, the development of counting rods, or 'Suanpan', displayed remarkable ingenuity in manipulating numeric values and paved the way for future mechanical calculators.

Contrary to what many might believe, Chinese scholars had already formulated binary arithmetic by the 3rd century BCE, predating its appearance in Western civilization by over a millennium. This early conception of binary numbers and their manipulation showcased Chinese mathematicians' extraordinary foresight and analytical prowess.

Thean illustration of renowned polymath Zhang Heng's seismoscope offers a glimpse into China's pioneering spirit in automated data collection and analysis. This ingenious invention not only recorded seismic activity but also employed mechanical principles and celestial observations in its operation. The seismoscope stands as a symbol of China's forward-thinking approach to technology and serves as an early precursor to modern-day sensor devices.

Furthermore, the significance of the 'hanzi' writing system should not be overlooked in the context of Chinese computing history. Each character in 'hanzi' represents a morpheme, a linguistic unit carrying semantic meaning, which introduces an inherent level of abstraction akin to the logic employed in programming languages. The systematic organization and classification of these characters highlight the Chinese civilization's early inclination toward structuring information, emphasizing the foundational role of language in computational thought.

In contemplating the mechanisms of ancient Chinese computations, historians and technologists are continually uncovering connections between early practices and contemporary digital frameworks. The innovative strides made in the realm of computation during this era not only demonstrate the profundity of ancient Chinese intellect but also underscore the enduring impact of their technological legacy on modern society.

Legacy and Influence: The Impact on Later Cultures

The legacy of advanced technology in ancient China resonates deeply within later cultures, leaving an indelible imprint on the course of human history. The innovations and achievements of ancient China continue to exert profound influence not only in Asia but across the globe. From the conceptualization of digital computations to the development of astronomical instruments, Chinese technology profoundly impacted subsequent civilizations in numerous ways.

One of the most significant legacies is the role of Chinese technological advancements in shaping global trade networks. Through their expertise in textile manufacturing, ceramics, papermaking, and other industries, ancient Chinese innovations enriched the pathways of commerce, facilitating the exchange of knowledge, goods, and cultural ideas. The Silk Road, for instance, stands as a testament to the enduring impact of ancient Chinese technology on international trade and interconnectedness.

Furthermore, the military ingenuity displayed by ancient China, such as the invention of gunpowder and the creation of powerful weapons, significantly influenced the development of warfare tactics and strategies in other societies. The impact of ancient Chinese military technology extended far beyond the battlefield, contributing to the evolution of defensive structures and fortifications around the world.

When considering the architectural and engineering marvels of ancient China, it becomes evident that these achievements established a lasting legacy, inspiring future generations of builders, engineers, and visionaries. The mastery of hydraulic engineering, including the construction of canals, dams, and irrigation systems, not only facilitated agricultural prosperity but also engendered admiration and emulation from later cultures seeking to harness the power of water for sustenance and development.

Moreover, the philosophical underpinnings and mathematical concepts that underpinned Chinese technological innovations have persisted throughout history, permeating diverse intellectual traditions and embodying enduring principles that continue to inform scientific thought and scholarly endeavors today. Concepts such as the harmonious relationship between heaven and earth, balance and equilibrium, and the relentless pursuit of precision have left an indelible mark on later cultures and continue to shape contemporary perspectives on science and technology.

In conclusion, the remarkable technological achievements of ancient China have reverberated through time, leaving an enduring legacy that has transcended geographical boundaries and cultural divides. The impact of these innovations on later cultures remains unmistakable, underscoring the profound and multifaceted influence of ancient Chinese technology on the trajectory of human civilization.

7

The Technological Marvels of Ancient Persia

Persian Innovations

Ancient Persia, renowned for its ingenuity and technological advancements, stands as a testament to the innovative spirit of human civilization. From the intricate water management systems to the awe-inspiring architectural marvels, Persian innovations illustrate the remarkable capabilities of ancient engineers and craftsmen.

At the heart of Persian innovation lay a deep understanding of hydraulic engineering, particularly exemplified by their sophisticated qanat system. These subterranean aqueducts, known for their exceptional design and construction, showcased the Persians' mastery in harnessing natural resources. The profound impact of these water management systems extended beyond providing vital irrigation for agricultural lands; they also facilitated the development of thriving urban centers and sustained the growth of the Persian Empire.

Moreover, the Persians excelled in military engineering, devising advanced fortifications and siege warfare techniques that were unrivaled in their time. Their expertise in architecture is evident in the majestic palaces and paradises that adorned the Persian landscape, each structure a testament to the fusion of artistic elegance and practical functionality. The Royal Road, an engineering feat that connected distant regions of ancient Persia, reflects the empire's commitment to efficient communication and transportation.

In the realm of scientific pursuits, Persian contributions to astronomy and navigation were notable, as scholars delved into the mysteries of the stars for celestial guidance and maritime exploration. Furthermore, the Persians were revered for their exceptional craftsmanship, particularly in metalworking and jewelry, producing ornate artifacts that reflected their cultural sophistication and mastery of metallurgy.

The intertwining of technological prowess with religious and philosophical ideologies, such as Zoroastrianism, played a significant role in shaping Persian innovations. It fostered a holistic approach to technological development, imbuing it with spiritual significance and a deeper connection to societal values and beliefs.

The enduring legacy of Persian innovations continues to resonate through the annals of history, influencing subsequent civilizations and inspiring modern-day technological advancements. By exploring the multifaceted dimensions of Persian ingenuity, we gain valuable insights into the interplay between culture, engineering, and innovation, illuminating the extraordinary achievements of the ancient Persian civilization.

Water Management Systems: Qanats and Beyond

Ancient Persia, known for its ingenuity and advanced technological marvels, showcased remarkable expertise in the realm of water management systems. At the heart of this expertise lay the intricate network of qanats, a subterranean aqueduct system that revolutionized Persia's approach to water supply. The qanat system, also referred to as 'kariz' in Persian, allowed for the efficient extraction and distribution of groundwater over vast distances, making agriculture sustainable and enabling urban development in arid regions. The qanat technology represented a sophisticated application of engineering and hydrology that remains awe-inspiring even by modern standards. Moreover, Persians pushed the boundaries of water management with other innovative solutions that extended beyond the qanat system. These included reservoirs, dams, and canals, demonstrating their comprehensive understanding of hydraulic engineering. The enduring impact of these ancient advancements can still be observed in the region's landscape, as remnants of these structures continue to stand as a testament to the Persians' mastery of water management. This exceptional feat of engineering further illustrates the profound influence that ancient Persia had on shaping sustainable habitats in challenging environmental conditions, setting a precedent for future civilizations in combating water scarcity. As we delve deeper into the intricacies of Persia's water management systems, it becomes evident that their technological prowess was not confined to land-based innovations alone. The application of such advanced systems also facilitated the flourishing of ancient Persia's trade and economy, contributing to its position as a thriving cultural and economic hub in the ancient world. Through the study of these historic water management practices, we gain valuable insights into the wisdom and resourcefulness of ancient civilizations, offering timeless lessons for sustainable development in the present day.

Military Might and Engineering: The Art of War

In ancient Persia, military might was a crucial aspect of the empire's power and influence. The Persian Empire, known for its formidable military prowess, displayed an unparalleled understanding of engineering in the context of warfare. This section delves into the intricate strategies and innovations deployed by the Persian forces, underscoring the profound impact they had on the art of war during this era.

One of the most remarkable displays of Persian military engineering was evident in their innovative use of siege tactics and weaponry. The development and deployment of advanced war machines such as siege towers, battering rams, and mobile shelters demonstrated the Persians' expertise in implementing sophisticated technologies to gain strategic advantages on the battlefield. Additionally, their utilization of expansive networks of fortifications and defensive structures showcased an exceptional understanding of defensive warfare.

Moreover, the engineering marvels of the Achaemenid army, including the construction of infrastructure for rapid mobilization and supply logistics, stand testament to the ingenuity of ancient Persian military engineers. The invention of efficient transportation systems to move troops, supplies, and equipment across vast territories bolstered the empire's ability to maintain and expand its hegemony, enabling swift response to incursions and revolts.

The Persian mastery of engineering extended beyond traditional warfare, encompassing naval and maritime innovation. The development of advanced shipbuilding techniques, coupled with astute navigational knowledge, allowed the Persians to establish dominance over vital waterways, thus securing critical trade routes and projecting authority across distant lands. This demonstrated a profound understanding of the importance of sea power in maintaining a comprehensive military strategy.

Understanding the technological and engineering advancements in ancient Persia sheds light on the comprehensive approach to warfare embraced by the empire. By engaging with the multifaceted nature of these developments, we can gain a deeper appreciation of the role that technology played in shaping the ancient world's military conquests and defense strategies.

Architectural Brilliance: Palaces and Paradises

In the annals of architectural history, the palaces and paradises of ancient Persia stand as a testament to the grandeur and sophistication of the Achaemenid and Sassanian empires. The Persian architects demonstrated their mastery in creating opulent and awe-inspiring structures that combined functionality with exquisite design. The palaces, such as the magnificent Persepolis, showcased intricate carvings, majestic staircases, and expansive courtyards that reflected the power and wealth of the empire. These architectural marvels were not only symbols of prestige but also served as administrative centers and ceremonial spaces. The innovative use of columned halls, audience chambers, and royal reception areas within these palaces exemplified the advanced understanding of spatial organization and acoustics. One cannot help but marvel at the ingenuity displayed in the layout and construction of these edifices, demonstrating an unparalleled level of architectural sophistication. Moreover, the paradises or gardens surrounding these palaces were meticulously designed, featuring flowing water channels, lush greenery, and vibrant flora, creating serene and picturesque retreats fit for kings and nobility. The renowned Hanging Gardens of Babylon, often attributed to Persian influence, were celebrated for their elevated terraces and complex irrigation systems, showcasing the mastery of landscape architecture. The architectural brilliance of these palaces and paradises not only impressed contemporaries but also influenced future civilizations, leaving an enduring legacy on architectural design and landscaping techniques. Understanding the innovations and technical prowess employed in constructing these ancient wonders provides profound insights into the rich cultural heritage and technological advancements of ancient Persia.

Roads and Communication: The Royal Road

The Royal Road, an extraordinary feat of engineering in ancient Persia, served as a pivotal thoroughfare connecting the vast Persian Empire. Stretching over 1,600 miles from Susa in the south to Sardis in the west, this expansive roadway facilitated efficient communication, transportation, and trade across diverse terrains and regions. Commissioned by King Darius I, the Royal Road symbolized the empire's prowess in infrastructure and logistics, enhancing both administrative control and cultural exchange. Its strategic significance was unrivaled, solidifying Persia's position as a dominant force in the ancient world.

Constructed with meticulous planning and innovative techniques, the road incorporated way stations, called caravanserais, strategically placed at regular intervals. These facilities provided essential amenities for travelers and merchants, ensuring safe passage and enabling the swift movement of goods and information. Additionally, relays of mounted couriers, known as Angarium, proficiently carried messages between key points along the route, reducing communication times significantly.

Furthermore, the Royal Road boasted advanced road-building methods, utilizing standardized measures for construction that resulted in a remarkably durable and well-maintained thoroughfare. The route traversed deserts, mountains, and plains, showcasing Persia's adeptness in overcoming geographical challenges. The development of the road also stimulated economic growth by fostering increased trade and cultural interchange among diverse populations.

This remarkable artery of communication and commerce not only facilitated the integration of far-flung territories but also exemplified the Persians' dedication to technological advancement and strategic foresight. Its legacy endured beyond the ancient era, influencing subsequent empires and societies and paving the way for future innovations in transportation and communication.

Astronomy and Navigation: Stars as Guides

In ancient Persia, the study of astronomy and navigation held profound significance in both scientific and philosophical realms. With a keen observation of the celestial bodies, Persian scholars made remarkable strides in understanding the movements of stars and planets, devising sophisticated astronomical instruments, and laying the groundwork for advanced methods of navigation. The night sky became a vast canvas against which early Persian astronomers charted their course, utilizing the stars as faithful guides across land and sea.

The Persians meticulously documented the positions of constellations, recognizing patterns that helped them predict seasonal changes and anticipate celestial events. They also observed the movement of prominent stars such as Sirius, developing a rudimentary calendar based on these celestial rhythms. Their adeptness in celestial observations fostered a comprehensive understanding of the intricate connection between the heavens and life on earth.

Beyond mere observation, ancient Persian navigators applied their astronomical knowledge to map out trade routes, exploring distant lands and establishing lucrative commercial networks. By aligning their navigational techniques with the positions of stars, they were able to embark on extensive journeys with confidence and precision. The mastery of celestial navigation not only facilitated trade and diplomacy but also enriched the cultural exchange between Persia and its trading partners, leaving an indelible mark on the annals of global history.

Furthermore, the Persian advancements in astronomy and navigation are evident in their creation of innovative instruments like the astrolabe and quadrant, which enabled precise measurement of celestial angles and distances. These tools revolutionized the science of celestial navigation, paving the way for future developments in maritime exploration and cartography. Through the astrolabe, sailors and scholars could determine their exact position at sea, facilitating safer and more efficient maritime endeavors. Additionally, Persian astronomers contributed to the refinement of the zodiac system, offering insights into the cyclical nature of time and the cosmic order.

The enduring legacy of Persian astronomy and navigation reverberates through the corridors of history, illuminating the profound impact of their discoveries and innovations. From navigating across uncharted seas to unraveling the mysteries of the universe, the Persians encapsulated the spirit of exploration and discovery, forging pathways that continue to inspire and guide humanity towards new horizons.

Metalworking Mastery and Jewelry Craftsmanship

Ancient Persia was renowned for its mastery of metalworking and jewelry craftsmanship, skills that were honed over centuries and enriched by influences from neighboring regions. Persian artisans were adept in working with a variety of metals, including gold, silver, copper, and bronze, employing sophisticated techniques to create exquisite objects of beauty and function. The artisanal expertise extended beyond mere adornment; it was a manifestation of the empire's rich cultural heritage and its deep-rooted connection to various artistic traditions. Indeed, the development of metallurgy in Persia was indicative of the empire's ingenuity and advanced technological capabilities. The craft of metalworking flourished throughout Persia, with each region showcasing distinct styles and preferences, reflecting the diverse tapestry of the Persian Empire. Skilled craftsmen utilized a range of techniques such as casting, engraving, filigree, and granulation to fashion intricate jewelry, decorative objects, and ceremonial regalia. These masterpieces often incorporated precious gemstones, pearls, and enameling, exhibiting an unparalleled attention to detail and a keen understanding of aesthetics. Persian jewelry was not only a symbol of wealth and status but also a reflection of spiritual and cultural beliefs, serving as potent expressions of identity and belonging. Moreover, the intricately crafted metalwork served practical purposes as well, with items such as fine tableware, weapons, and architectural adornments demonstrating the sophistication and versatility of Persian metallurgy. The importance of metalworking extended beyond individual craftsmanship, influencing broader aspects of Persian society, economy, and trade. The finely wrought metal artifacts became coveted commodities, sought after by neighboring civilizations and distant empires alike, fostering networks of exchange and cultural interaction across vast distances. Persian metalwork encompassed a spectrum of forms, from delicate jewelry and ornate drinking vessels to monumental statues and majestic architectural elements, each exemplifying the consummate skill and creative vision of the artisans. This legacy continues to inspire contemporary artists and craftsmen, underscoring the enduring impact of ancient Persian metalworking and jewelry craftsmanship on the world's cultural heritage and artistic traditions. The artistry and technical prowess of Persian metalworkers have left an indelible mark on history, embodying the empire's commitment to innovation, creativity, and excellence.

Textiles and Dyes: Cultural Expressions

Textiles and dyes played a pivotal role in the cultural expressions of ancient Persia, serving as a testament to the sophistication and artistic prowess of the civilization. Persian textiles were renowned for their intricacy, luxuriousness, and vibrant colors, reflecting the craftsmanship and sensibilities of the people. The weaving and dyeing techniques employed by Persian artisans were highly advanced for their time, enabling the creation of exquisite fabrics that found favor not only within the empire but also in distant lands through trade and diplomatic exchanges.

At the heart of Persian textile production was the art of carpet weaving, which remains a celebrated tradition to this day. The carpets woven by Persian artisans were distinguished by their intricate designs, fine knotting, and rich color palettes, showcasing the fusion of technical skill with creative expression. Each carpet served as a canvas for storytelling, with motifs and patterns often carrying symbolic meanings and historical narratives, further elevating their significance beyond mere utilitarian objects. Furthermore, the process of dyeing the wool or silk utilized in these carpets involved the use of natural pigments derived from plants, insects, and minerals, resulting in a stunning array of hues that represented the diverse landscapes and cultures of Persia.

In addition to carpets, Persian textiles encompassed a wide range of materials, including silks, linens, and cotton, each bearing distinct characteristics and embellishments tailored to various purposes and social statuses. The royal courts of Persia particularly favored sumptuous silk textiles adorned with intricate embroideries, gold threadwork, and elaborate patterns, signifying opulence, refinement, and status. On the other hand, simpler yet elegant garments, such as tunics and shawls, were crafted for everyday wear, reflecting the fusion of comfort and aesthetics in Persian attire.

The significance of textiles extended beyond mere ornamentation, encapsulating the essence of Persian culture and identity. These fabric creations embodied the interplay of tradition, religion, and artistry, with certain motifs and colors holding deep symbolic meanings rooted in Zoroastrian beliefs and mythological narratives. Furthermore, the excellence of Persian textiles transcended domestic boundaries, gaining recognition and admiration across the ancient world, thereby influencing the development of textile arts in neighboring regions and civilizations.

Ultimately, the mastery of textile production and dyeing in ancient Persia stands as a testament to the ingenuity, cultural richness, and technological advancement of the civilization. Through the splendor of their textiles, the Persians wove a tapestry that reflected the diversity, creativity, and enduring legacy of their society, leaving an indelible mark on the annals of human artistic achievement.

The Influence of Zoroastrianism on Technological Development

Zoroastrianism, one of the oldest known religions, played a significant role in shaping the technological advances of ancient Persia. The core tenets of this faith, including the pursuit of purity, truth, and cosmic order, instilled a deep respect for nature and the environment. This reverence for the natural world had a profound impact on the development of various technologies in Persian society. One of the fundamental principles of Zoroastrianism is the concept of 'asha' or righteousness, which encompassed ethical, social, and environmental responsibility. As a result, adherents of this faith were inclined to develop technologies that promoted harmony with nature and upheld moral integrity. This ethos influenced advancements in agriculture, architecture, and water management. The reverence for purity and cleanliness in Zoroastrianism also led to the creation of advanced sanitation systems and techniques for maintaining hygiene. Furthermore, the emphasis on cosmic balance within Zoroastrian teachings spurred the development of precise astronomical instruments and contributed to advancements in celestial navigation. The religious focus on the duality between good and evil encouraged early Persian scholars to delve into the study of metallurgy, seeking to extract and refine metals for virtuous purposes. This pursuit laid the foundation for intricate metalworking techniques, enabling the creation of exquisite jewelry and ornamental designs. Moreover, the Zoroastrian emphasis on community welfare and righteous governance fostered advancements in urban planning, leading to the construction of grand cities and efficient infrastructure. Zoroastrianism's influence extended beyond tangible technologies, permeating into the realm of intellectual endeavors. The quest for truth and wisdom, central to the faith, resulted in the establishment of academies and libraries that became centers of learning and innovation. Zoroastrian scholars made significant contributions to the fields of medicine, mathematics, and philosophy, laying the groundwork for future advancements in these areas. Overall, the amalgamation of Zoroastrian beliefs and technological progress created a distinctive legacy in ancient Persia, leaving an indelible mark on the trajectory of global civilization.

Legacy: Persian Contributions to World Technology

The legacy of ancient Persia in shaping world technology is substantial and multifaceted. Persian contributions have left an indelible mark on various domains, fostering advancements that have echoed through history. One of the most noteworthy legacies is the establishment of a sophisticated postal system, known as the Chapar Khaneh, which played a crucial role in facilitating communication and trade across vast distances. This innovation influenced subsequent postal systems and logistics networks, exemplifying Persian prowess in enhancing global connectivity.

Furthermore, Persian advancements in irrigation and water management, with the ingenious qanat system leading the way, showcased a deep understanding of hydro-engineering that revolutionized agricultural practices not only in Persia but also in regions they influenced.

This legacy lives on in modern water management techniques, underscoring the enduring impact of ancient Persian wisdom.

Persian contributions to architecture, particularly exemplified by the awe-inspiring palaces and paradises such as Persepolis, introduced innovative construction methods and design principles that continue to inspire architects and engineers worldwide. The intricate artistry and engineering marvels found in these structures reflect a commitment to excellence that has earned them a place as timeless symbols of human achievement. Moreover, the Royal Road, a network of well-maintained highways, facilitated trade and cultural exchange, setting a precedent for the development of efficient transportation infrastructures that transcended borders and united diverse civilizations.

In addition, Persian expertise in astronomy and navigation was instrumental in advancing celestial studies and maritime exploration. Navigational tools and astronomical knowledge developed in Persia greatly enhanced global seafaring efforts, contributing to the collective pool of human knowledge and furthering navigation technologies. Persian metallurgy and jewelry craftsmanship, infused with artistic and technical finesse, set standards that influenced metalworking traditions across continents, leaving enduring imprints on cultural and technological landscapes.

Persian textiles and dyes, renowned for their richness and quality, not only adorned the elite but also permeated distant markets, serving as a testament to Persian innovations in textile manufacturing and dyeing processes. The pervasive influence of Zoroastrianism on technological development cannot be overlooked, as it motivated ethical and philosophical considerations that drove advancements across disciplines, infusing an unparalleled spiritual dimension into Persian ingenuity.

Ultimately, the lasting legacy of Persian contributions to world technology extends beyond individual achievements, encompassing a collective ethos of progress and innovation. The enduring resonance of these contributions underscores the profound impact of ancient

Persian civilization on shaping the course of human technological advancement.

8

The Ingenious Innovations of Ancient Greece

Introduction to Greek Technological Mastery

Greek technological mastery stands as an enduring testament to the creative ingenuity and engineering prowess of ancient Hellenic civilization. From architectural marvels to pioneering mechanisms, the innovative spirit of the Greeks permeated a vast array of domains, leaving an indelible mark on human history. The legacy of Greek technological advances continues to resonate in contemporary engineering principles, serving as a source of inspiration and a touchstone for modern innovation. The breadth and depth of their achievements exemplify an unparalleled dedication to precision, functionality, and aesthetic grandeur, propelling Greek technological mastery into an unparalleled realm of influence. Indeed, the impact of these advances extends far beyond their historical origins, offering timeless lessons that continue to shape our understanding of engineering, design, and problem-solving. By delving into the multifaceted landscape of Greek technological mastery, we uncover a rich tapestry of brilliance that has transcended time and space, casting a brilliant light upon the enduring power of human creativity and intellect.

Architectural Achievements: Beyond Aesthetics

The architectural achievements of ancient Greece are renowned for their remarkable blend of functionality and aesthetic beauty. Greek architecture represents the intersection of art, engineering, and cultural expression, leaving an indelible mark on the history of civilization. The structural innovations of the Greeks not only defined their built environment but also permeated their society, reflecting their ideals and aspirations. At the core of Greek architectural prowess lay a deep understanding of geometry, materials, and structural integrity. From the stately elegance of the Parthenon to the practical functionality of the Greek theatre, these structures were designed with meticulous attention to detail and purpose. The Doric, Ionic, and Corinthian orders exemplify the refinement and sophistication that characterized Greek architectural styles, influencing countless generations of architects and craftsmen. Beyond their stunning visual appeal, these architectural wonders served practical functions that shaped the daily lives and communal experiences of ancient Greek citizens. Notably, the incorporation of sophisticated water management systems, such as aqueducts and fountains, showcased the Greeks' ingenuity in harnessing natural resources for both utilitarian and ornamental purposes. Moreover, the strategic placement and design of buildings within city-states reflected the values and priorities of Greek society, fostering civic engagement and public assembly while enhancing the cohesion of urban spaces. It is essential to acknowledge that Greek architecture was not solely a feat of construction but also an enduring testament to their cultural identity and visionary spirit. Through their architectural achievements, the ancient Greeks demonstrated a profound reverence for the harmony between humanity and the environment, laying the foundation for enduring principles that continue to resonate in contemporary architectural practices. The legacy of Greek architecture serves as a timeless source of inspiration, reminding us of the profound impact that innovative design and structural mastery can have on shaping our physical surroundings and cultural heritage.

The Antikythera Mechanism: An Ancient Computer

Dating back to the 2nd century BCE, the Antikythera Mechanism stands as a testament to the remarkable ingenuity of ancient Greek technology. Discovered in a shipwreck near the island of Antikythera, this extraordinary artifact has fascinated scholars and historians for its advanced mechanical design and astronomical capabilities. The device itself is a complex arrangement of gears and dials, offering insight into the unparalleled knowledge and understanding of celestial motions possessed by the ancient Greeks.

One of the most astonishing features of the Antikythera Mechanism is its ability to predict astronomical phenomena, such as eclipses and the positions of celestial bodies, with remarkable precision. This level of sophistication far exceeds what was previously believed to be achievable during that era, challenging our conventional perceptions of ancient technological prowess. As a navigational tool, this mechanism provided invaluable assistance to seafaring, aiding sailors in determining their position relative to the stars and planets, further emphasizing the significance of this innovation in the context of ancient maritime exploration.

The intricacy and complexity of the Antikythera Mechanism underscore the advanced state of craftsmanship and technical knowledge present in ancient Greek society. Its existence raises compelling questions about the sources of such expertise and the extent to which it may have been disseminated across cultures. Furthermore, the implications of this remarkable artifact extend beyond the realm of technology and science, shedding light on the interconnected nature of ancient civilizations and the exchange of ideas and innovations.

In contemporary times, the Antikythera Mechanism continues to captivate and inspire researchers, serving as a symbol of the enduring legacy of ancient Greek technological achievements. Through ongoing study and analysis, scholars strive to unravel the mysteries surrounding this extraordinary device, seeking to gain deeper insights into the intellectual prowess and inventive spirit of the ancient world. The Antikythera Mechanism transcends its temporal confines, standing as a beacon of human curiosity, creativity, and the relentless pursuit of knowledge.

Military Machines: Enhancing Warfare Capabilities

The innovative prowess of ancient Greece extended far beyond the realms of art and philosophy; it also manifested in the development of advanced military machines that revolutionized warfare capabilities. Greek engineers and inventors demonstrated exceptional skill and ingenuity in creating formidable weapons and defensive mechanisms that played a pivotal role in shaping the outcomes of battles and conflicts. The use of ingenious military technology not only provided tactical advantages but also influenced the strategic dynamics of ancient warfare. One of the most iconic innovations was the development of powerful war machines such as the 'Polybolos,' a repeating ballista capable of launching multiple projectiles with remarkable precision and force. Its ability to unleash a barrage of projectiles at a rapid pace bestowed significant superiority upon the Greek forces, changing the course of many confrontations. Moreover, the creation of more sophisticated siege engines like the 'Torsion Catapult' demonstrated the Greeks' unparalleled expertise in employing mechanical power for destructive ends, enabling them to breach fortified walls and structures with unprecedented efficiency. The deployment of such cutting-edge military machinery showcased the Greek commitment to continuous technological improvement and their dedication to achieving superiority on the battlefield. In addition to offensive capabilities, the Greeks also excelled in crafting intricate defensive fortifications and constructions that bolstered their stronghold positions. Advanced city walls, fortified gateways, and strategically positioned watchtowers exemplified their mastery of architectural and engineering principles in the context of military defensive strategies. These innovations not only provided physical protection but also served as psychological deterrents, deterring potential adversaries from engaging in direct assaults. Furthermore, the utilization of naval technologies, such as the 'Trireme' warship, highlighted the Greeks' proficiency in maritime warfare, allowing them to project power across the Mediterranean and assert dominance over crucial trade routes. The precision and agility of these warships, equipped with bronze battering rams and multiple layers of oars, represented a significant leap forward in naval warfare tactics. The integration of advanced military machines and naval technologies into Greek warfare exemplifies the profound impact of technological innovation on ancient battlefields, reshaping the dynamics of conflicts and influencing the course of history. The legacy of these ingenious military innovations extends far beyond their immediate tactical applications, leaving an indelible mark on the evolution of warfare and the enduring influence of technological advancements in human endeavors.

Hydraulic Engineering: Mastery of Water

Ancient Greece was not only known for its advancements in philosophy, art, and warfare but also for its remarkable feats in hydraulic engineering. The Greeks exhibited a profound understanding of the principles governing water and its behavior, which allowed them to masterfully manipulate this vital resource for various purposes. One of the most iconic examples of their hydraulic expertise is the invention and implementation of aqueducts and irrigation systems. The construction of aqueducts played a pivotal role in supplying water to cities, ensuring that urban populations had access to clean and reliable water sources. This innovation not only improved public health but also facilitated the growth and prosperity of Greek urban centers. Furthermore, the sophisticated irrigation methods developed by the Greeks greatly enhanced agricultural productivity, enabling the cultivation of crops in arid regions and promoting food security. The intricate network of channels and canals designed by Greek engineers effectively distributed water across farmlands, contributing to sustainable agricultural practices and economic development. Their mastery of hydraulic engineering extended beyond practical applications and had profound implications for the aesthetic and recreational aspects of ancient Greek society. Fountains, ornamental pools, and elaborate water features adorned public spaces, reflecting the Greeks' appreciation for the artistic potential of water manipulation. These captivating creations not only served as focal points of beauty and tranquility but also demonstrated the ingenuity of hydraulic design. Moreover, the Greeks recognized the potential for harnessing water's kinetic energy, laying the foundation for early water-powered devices such as water mills. These innovations revolutionized labor-intensive processes, particularly in grinding grain and processing materials, thereby enhancing efficiency and productivity in various industries. The influence of Greek hydraulic engineering reverberated throughout the Mediterranean world, leaving an indelible mark on subsequent civilizations and inspiring further advancements in water management and infrastructure. The enduring legacy of their mastery of water serves as a testament to the profound impact that ancient Greek technological ingenuity had on shaping the course of history.

Philosophical Foundations of Greek Innovation

In ancient Greece, the foundations of innovation were deeply rooted in the philosophical underpinnings of the time. Philosophers such as Plato and Aristotle laid the groundwork for critical thinking, logic, and the pursuit of knowledge that directly influenced technological advancements. The concept of 'techne,' which encompassed both art and craftsmanship, was integral to Greek philosophy, emphasizing the connection between intellectual pursuits and practical applications. This philosophical perspective facilitated a holistic approach to innovation, where theoretical understanding was intricately woven with practical implementation. Greek philosophers' emphasis on rational inquiry and empirical observation spurred the development of sophisticated technological solutions across various disciplines. Furthermore, the cultivation of a questioning mindset and a thirst for knowledge led to the exploration of natural phenomena and the development of scientific principles.

The philosophical pillars of ancient Greece also championed the powerful notion of human agency and ingenuity. The belief in the inherent capabilities of humans to shape and transform their environment cultivated a culture of experimentation and bold innovation. It fueled the drive to harness natural resources, improve existing technologies, and envision groundbreaking inventions. Moreover, the philosophical discourse surrounding ethics and the greater good fostered a sense of responsibility in the application of technology. Ethical considerations became intrinsic to the innovation process, leading to the creation of technologies that served not only functional purposes but also upheld moral and societal values. This ethical dimension continues to resonate in contemporary discussions on the responsible use of technology.

Additionally, the interplay between Greek philosophy and innovation extended beyond mere practicalities; it encapsulated an overarching worldview that celebrated the unity of the cosmos, the pursuit of harmony, and the interconnectedness of nature and humanity. This harmonious paradigm inspired technological endeavors that sought to align with natural rhythms and laws, resulting in sustainable and enduring solutions. The amalgamation of philosophical ideals with technical prowess not only manifested in remarkable innovations but also left an indelible mark on subsequent civilizations and the ongoing evolution of human progress.

Ultimately, the philosophical foundations of Greek innovation signify an intrinsic union of wisdom, creativity, and human purpose, illuminating a profound understanding that transcends the mere material aspect of technology. They encapsulate a timeless ethos that reverberates through the annals of history and serves as a testament to the profound impact of philosophical inquiry on the tapestry of human ingenuity and progress.

Material Science and Metallurgy

The ancient Greeks were renowned not only for their intellectual prowess and philosophical insights but also for their profound contributions to the realm of material science and metallurgy. This chapter delves into the intricate craftsmanship and technological advancements that underpinned Greek society, illustrating their far-reaching impact on both artisanal and industrial domains. Material science, as understood in the context of ancient Greece, encompassed a broad spectrum ranging from the development of innovative construction materials to the manipulation and refinement of metals such as bronze, iron, and silver. The mastery of metalworking techniques facilitated the production of a diverse range of artifacts, from exquisitely crafted jewelry and ornamental objects to practical tools and weapons. Notably, the Greeks exhibited unparalleled skill in forging bronze, a pivotal metal in their cultural and technological landscape. Through an in-depth exploration of alloying, casting, and metal finishing, the Greeks elevated metallurgical craftsmanship to unprecedented heights, setting new benchmarks for durability, aesthetic appeal, and functional sophistication. Moreover, their adept understanding of material properties fostered significant advancements in architectural engineering, enabling the construction of monumental structures adorned with intricately detailed metal embellishments. The significance of material science and metallurgy extended beyond artistic expression, permeating into vital sectors such as coinage, where standardized metallic currency served as a hallmark of economic prosperity and administrative efficacy. Furthermore, the qualitative analysis and experimentation undertaken by ancient Greek metallurgists reverberated throughout subsequent centuries, enriching the foundations of modern scientific inquiry and establishing enduring principles of empirical observation and deductive reasoning. As this elucidates, the legacy of Greek material science and metallurgy persists as a testament to their innovation, adaptability, and unwavering pursuit of knowledge.

Influence on Mediterranean Trade and Economy

The innovative advancements in material science and metallurgy achieved by the ancient Greeks had a profound influence on the trade and economy of the Mediterranean region. Greek expertise in crafting exceptional quality metalwork, such as bronze sculptures, elegantly decorated vases, and finely wrought jewelry, became highly sought after throughout the Mediterranean world. This demand not only facilitated extensive trading networks but also significantly contributed to the economic prosperity of Greek city-states while further integrating them into the broader regional economy.

The exportation of Greek metalwork had a substantial impact on shaping regional trade dynamics. Goods crafted from precious metals, intricate metalworking tools and techniques, and knowledge of alloy composition were eagerly traded across the Mediterranean. As a result, Greek city-states discovered new opportunities for commercial exchange and established strong economic ties with neighboring cultures and civilizations. These interactions not only facilitated the flow of goods but also enabled the exchange of ideas, philosophies, and technological innovations, thereby fostering cultural interconnectedness throughout the region.

The burgeoning market for Greek metalwork and metallurgical products also ignited further specialization and diversification within local economies, driving the need for skilled artisans, merchants, and intermediaries to facilitate trade routes. The high value placed on Greek metalwork not only boosted the wealth of individual craftsmen and traders but also enriched the coffers of the city-states that sponsored their work. Furthermore, the economic stimulus generated from the dissemination of Greek metallurgical products helped fuel urbanization, supported infrastructure development, and provided financial means for endeavors in art, literature, and civic institutions.

Moreover, the exchange of valuable materials and artisanal expertise through cross-cultural trade fostered an environment conducive to the proliferation of innovation and technological know-how. The introduction of new manufacturing methods, design influences, and raw materials from distant regions significantly augmented the creative potential of Greek craftsmen, resulting in the birth of unique and coveted products that upheld the ingenuity and excellence synonymous with Greek craftsmanship.

As Greek metallurgy continued to command a prominent position in the Mediterranean economy, the far-reaching effects of this influence can be seen in the historical record. Notably, the enduring legacy of Greek metalwork and metallurgical prowess is reflected in the archaeological remains and artifacts discovered at flourishing trading centers, which serve as testaments to the expansive economic networks facilitated by Greek technological contributions. The enduring impact of Greek material science and metallurgical expertise reverberates through the annals of history, underscoring the pivotal role played by ancient Greek innovations in shaping the landscape of Mediterranean trade and economy.

Cross-Cultural Exchanges and Technology Adaption

Cross-cultural exchanges played a pivotal role in the dissemination and adaptation of Greek technological innovations throughout the ancient world. As the Greek civilization interacted with other cultures around the Mediterranean, a rich tapestry of knowledge exchange and technological transfer emerged, ushering in an era of unprecedented progress and transformation.

The Greek colonies established across the Mediterranean basin became focal points for the amalgamation of diverse cultural and technological influences. These outposts served as hubs of trade, commerce, and intellectual exchange, facilitating the diffusion of Greek advancements to distant shores and fostering a dynamic interplay of ideas and techniques. Through these interactions, Greek innovations in fields such as architecture, engineering, and mechanics were transmitted to regions as far-reaching as Egypt, Mesopotamia, and beyond.

One notable example of cross-cultural technological adaption is evident in the architectural styles and construction techniques observed in structures across the Mediterranean. The enduring influence of Greek architectural principles, including the implementation of columns, entablatures, and proportional design, reverberated throughout the region, leaving an indelible imprint on the built environment of diverse cultures.

Furthermore, the integration of Greek hydraulic engineering practices, such as aqueducts and water management systems, greatly impacted the development of urban infrastructure in neighboring civilizations. These innovative methods not only enhanced the efficiency of resource utilization but also facilitated the establishment of flourishing cities and agricultural landscapes in regions influenced by Greek expertise.

In the realm of material science and metallurgy, the assimilation of Greek knowledge and expertise catalyzed advancements in metalworking, resulting in the production of superior weaponry, tools, and artifacts. The transference of Greek metalworking techniques and alloy formulations contributed to the proliferation of high-quality metallic objects and heralded a new era of craftsmanship and industrial capability.

Moreover, the philosophical underpinnings of Greek innovation permeated diverse societies, shaping modes of critical thinking and problem-solving across different cultural contexts. The emphasis on observation, rational inquiry, and empirical experimentation espoused by Greek scholars engendered a paradigm shift in the approach to scientific exploration and technological development, influencing the intellectual landscape of cultures that intersected with Hellenic thought.

Ultimately, the cross-cultural diffusion and adaptation of Greek technological innovations not only accelerated the pace of progress but also fostered a climate of collaboration and mutual enrichment. This vibrant exchange of knowledge and expertise laid the groundwork for the interconnected web of ancient civilizations, cultivating a legacy of technological diversity and collective ingenuity that continues to resonate through the annals of history.

Conclusion: Legacy of Greek Innovations

The legacy of ancient Greek innovations resonates through the annals of human history, casting a profound and enduring influence on diverse fields. The cross-cultural exchanges and adaptive technologies emanating from the Hellenic world have left an indelible mark, shaping the trajectory of technological advancement for generations to come. The visionary ethos of Greek ingenuity permeates modern society in manifold ways, underscoring the pervasive and timeless impact of their innovations. The multifaceted legacy of Greek technological prowess encompasses architectural marvels, groundbreaking scientific advancements, pioneering philosophical concepts, and unparalleled artistic achievements, engendering an enduring imprint on the tapestry of human progress. The indomitable spirit of innovation that characterized the ancient Greeks continues to serve as a wellspring of inspiration, fueling contemporary endeavors at the vanguard of technology and knowledge. From the enduring principles of democracy to the enduring influence of Greek mathematics and astronomy, the enduring legacy of Greek innovations serves as a testament to the enduring power of human creativity and intellect.

One of the most profound conduits through which the legacy of Greek innovations endures is in the realms of architecture and engineering. The grandeur of ancient Greek structures such as the Parthenon and the Temple of Olympian Zeus stands as a testament to their astonishing architectural mastery. Their innovative use of columns, entablatures, and friezes not only epitomizes their advanced understanding of engineering but also artistically reflects their adroit harmonization of form and function. Furthermore, the revolutionary contributions of Greek hydraulic engineers, manifested in ingenious aqueducts, irrigation systems, and water mills, showcase their unparalleled expertise in harnessing the power of water, laying the foundation for future developments in hydraulic engineering.

Moreover, the trailblazing Antikythera Mechanism encapsulates the pinnacle of ancient Greek technological sophistication, serving as an exemplar of their profound comprehension of intricate mechanisms and astronomical phenomena, a harbinger of the scientific revolution that would later unfold. The military stratagems and innovations conceived by the Greeks, ranging from advanced warships to complex siege engines, reflect their commitment to pushing the boundaries of warfare technology, thereby reshaping the dynamics of armed conflict and fortification design.

The inestimable value of Greek philosophic tenets, which underpin fundamental ideals such as rational inquiry, empirical observation, and critical analysis, continues to permeate myriad domains of modern thought. From the foundational principles of Western philosophy to the enduring legacy of ethical theories, the intellectual lineage of the Greeks reverberates throughout the disciplines of ethics, epistemology, metaphysics, and political theory. Their profound corpus of intellectual inquiries serves as a cornerstone of modern science, mathematics, and scholarly discourse, perpetuating an enduring legacy that transcends temporal and cultural boundaries.

In conclusion, the legacy of ancient Greek innovations stands as an enduring testament to the unyielding spirit of human ambition, collaboration, and discovery. The multifaceted contributions of the ancient Greeks reverberate across the ages, inspiring contemporary generations to embark on extraordinary quests for knowledge, creativity, and technological innovation. The enduring legacy of Greek innovations serves as a perennial source of admiration, reverence, and emulation, propelling humanity towards ever-greater heights of achievement and enlightenment.

9

Ancient Engineering in the Americas

Unveiling Mesoamerican Ingenuity

The allure of Mesoamerican civilization has fueled scholarly curiosity and public intrigue for centuries. From the first encounters between European explorers and the enigmatic ruins of ancient cities to contemporary archaeological excavations, the mystique of Mesoamerica's technological achievements continues to captivate minds around the globe. The complexity of Mesoamerican societies, their architectural marvels, intricate artistry, and sophisticated mathematical and astronomical systems have all contributed to a rich tapestry of cultural heritage that beckons us to unravel its mysteries. The fascination with Mesoamerican ingenuity is not limited to academic circles but extends to popular culture, inspiring novels, films, and exhibitions that seek to bring this extraordinary civilization to life. This chapter sets the stage for a comprehensive exploration of Mesoamerican technological advancements by delving into the geographical and cultural contexts that provided the fertile ground for such remarkable innovations. By gaining a deeper understanding of Mesoamerica's unique environment and the societal dynamics that shaped its development, we can better appreciate the profound impact of this ancient civilization on the history of human achievement.

Geographical Overview and Cultural Context

The Mesoamerican region is a diverse and remarkable landscape that served as the fertile ground for the development of ancient civilizations such as the Olmec, Maya, Aztec, and Zapotec. This area encompasses present-day Mexico, Belize, Guatemala, El Salvador, Honduras, Nicaragua, and Costa Rica, offering a rich tapestry of geographical features ranging from dense rainforests to high plateaus and coastal plains. The varying topography influenced the settlement patterns and agricultural practices of these ancient peoples, leading to the emergence of thriving city-states with distinctive cultural identities. The tropical climate and abundant natural resources supported the growth of complex societies, enabling the Mesoamericans to engage in trade, craft intricate artworks, and develop sophisticated architectural and engineering feats.

The cultural context of Mesoamerica is characterized by its unique religious beliefs, elaborate cosmology, and ritual practices that were intricately interwoven with everyday life. The Mesoamericans revered a pantheon of deities representing natural forces and celestial bodies, influencing their social structure, political organization, and artistic expressions. The cities served as centers of power and worship, adorned with elaborate temples, pyramids, and palaces that reflected the spiritual and ideological foundations of these ancient societies. Furthermore, the complex social hierarchies and extensive trade networks fostered cultural exchange and the transmission of knowledge, contributing to the flourishing of intellectual pursuits and technological innovations. The profound impact of environment and spirituality shaped the Mesoamerican worldview, providing a compelling backdrop for understanding the remarkable advancements achieved by these enigmatic civilizations.

As we delve into the geographical overview and cultural context of Mesoamerica, it becomes evident that the interplay between natural surroundings, religious beliefs, and societal dynamics profoundly influenced the trajectory of technological progress and artistic creativity in this region. By exploring the intricate connections between geography, culture, and civilization, we can gain a deeper appreciation for the ingenuity and resilience exhibited by the inhabitants of Mesoamerica, whose enduring legacy continues to captivate and inspire us today.

Architectural Achievements: Pyramids and Cities

The architectural prowess of Mesoamerican civilizations is epitomized in their stunning pyramids and intricately designed cities. These structures stand as testaments to the engineering ingenuity and cultural sophistication of ancient Mesoamerica. The iconic Mesoamerican pyramids, such as the Pyramid of the Sun in Teotihuacan and the Temple of Kukulcan at Chichen Itza, showcase not only the skillful construction techniques but also the deep spiritual and cosmological significance embedded within their design. These monumental edifices were often aligned with celestial events and served as grandiose platforms for religious ceremonies and societal gatherings. Their construction involved complex planning, precise measurements, and an understanding of astronomical alignments that reflected the Mesoamerican people's advanced knowledge of mathematics and astronomy.

In addition to the pyramids, the cities of Mesoamerica exhibited remarkable urban planning and architectural finesse. The layout of these cities was meticulously organized, featuring monumental plazas, ball courts, and residential areas. Teotihuacan, known for its broad avenues and vast ceremonial spaces, exemplifies the urban sophistication achieved by the ancient Mesoamerican civilization. The city's grid layout and monumental architecture reveal a high level of centralized planning and communal organization, reflecting the degree of social and political complexity within Mesoamerican societies. Moreover, the cities were adorned with intricate murals, sculptures, and artisanal works that depicted religious beliefs, historical narratives, and everyday life, showcasing the multifaceted artistic and cultural expressions of these ancient communities.

Furthermore, the architectural achievements of Mesoamerica extended beyond the construction of individual structures and urban centers. The development of advanced engineering techniques enabled the creation of extensive road networks and water management systems that facilitated trade, communication, and agricultural productivity. The control of water through aqueducts, canals, and reservoirs was crucial for sustaining agricultural activities in diverse environmental settings, showcasing the Mesoamerican civilization's mastery of environmental adaptation and resource utilization. Altogether, the architectural legacy of Mesoamerica reflects a harmonious integration of spiritual, astronomical, artistic, and practical elements, underscoring the holistic nature of technological advancements within this ancient culture.

Astronomical Excellence and Calendrical Systems

The Mesoamerican civilizations exhibited a profound understanding of astronomy and developed intricate calendrical systems that continue to intrigue and astonish modern scholars. Their sophisticated observations of celestial bodies allowed them to create calendars that were remarkably accurate in tracking the movements of the sun, moon, and planets. This deep comprehension of astronomical cycles was essential for agricultural planning, religious rituals, and governance, demonstrating the integral role of science in the cultural fabric of these ancient societies.

The Maya, in particular, were renowned for their comprehensive calendar systems. Their Long Count calendar, consisting of interlocking cycles, enabled them to record dates spanning thousands of years with remarkable precision. Moreover, the correlation between astronomical events and significant historical or mythological occurrences was a fundamental aspect of their worldview. The enigmatic alignment of Maya temples with celestial phenomena further attests to their advanced astronomical knowledge, indicating a purposeful integration of architecture and cosmology.

In addition to the Maya, other Mesoamerican cultures, such as the Aztecs and the Zapotecs, also developed complex calendars. These systems incorporated multiple interwoven cycles governing diverse aspects of life, including agricultural cycles, religious ceremonies, and political events. The careful observation of celestial motions and their correlation with terrestrial phenomenon allowed these civilizations to construct calendars that fostered a deeper understanding of the intertwined nature of cosmic and earthly processes.

Ethnoastronomy, the study of how cultures perceive and interpret celestial phenomena, has shed light on the profound significance of astronomy within Mesoamerican societies. The practice of aligning structures with solstices and equinoxes, as well as documenting eclipses and planetary movements, underscores the deep reverence for celestial bodies and their influence on human existence. By delving into the intricacies of Mesoamerican calendrics and their relationship to celestial observations, contemporary researchers continue to unravel the intellectual achievements of these ancient peoples, revealing a wealth of knowledge that extends far beyond mere timekeeping.

Furthermore, the enduring legacy of Mesoamerican calendrical systems is evident in contemporary society. The resilience and adaptability of these ancient calendars, particularly the Maya calendar, have transcended time, fostering a renewed fascination with their astronomical heritage. The enduring interest in Mesoamerican astronomy and calendrics serves as a testament to the enduring impact of these ancient technologies, inspiring ongoing inquiries into the intersection of science, culture, and spirituality.

The Role of Rituals and Religion in Technological Advancements

Religion and rituals played a pivotal role in shaping Mesoamerican technological advancements, serving as the driving force behind numerous innovations that emerged within these ancient civilizations. The sacred beliefs and complex rituals of societies such as the Maya, Aztec, and Zapotec were intrinsically linked to their understanding of the natural world, leading to the development of sophisticated technologies that embodied spiritual significance. At the core of Mesoamerican cultures, religion permeated every aspect of daily life, influencing the progression of agricultural techniques, urban planning, and astronomical knowledge. The seamless integration of religious practices with technological endeavors resulted in remarkable achievements and lasting legacies.

In examining the role of rituals in technological advancements, it becomes evident that these ancient societies meticulously aligned their innovations with spiritual concepts and celestial events. The construction of monumental temple pyramids and observatories reflected not only architectural prowess but also a deep reverence for deities and cosmological entities. Moreover, the intricate calendrical systems devised by Mesoamerican civilizations were intricately woven into religious ceremonies, demonstrating an advanced understanding of astronomy and timekeeping. The interconnectedness of rituals and technology was further manifested through the creation of complex ritual paraphernalia, such as ceremonial vessels and ornate sculptures, incorporating specialized artisanal skills and symbolic motifs rooted in religious narratives.

Furthermore, the influence of religious practices extended beyond material culture, spurring technological ingenuity in agriculture and resource management. Rituals associated with agricultural cycles and seasonal transitions drove innovative irrigation techniques and terraced farming methods, optimizing food production and sustainability. The spiritual significance attributed to natural elements and agricultural fertility fostered the implementation of sophisticated water systems, enabling efficient utilization of resources and providing vital sustenance for burgeoning Mesoamerican societies. This integral fusion of spirituality and practicality underpinned the evolution of advanced agrarian technologies.

The profound impact of rituals and religion on technological advancements is undeniable, profoundly shaping Mesoamerican civilization and leaving an enduring imprint on the development of innovative tools, engineering marvels, and societal organization. By understanding the intricate interplay between spiritual beliefs and technological progress, we gain valuable insights into the complexities of ancient societies and the multifaceted motivations driving their quest for advancement.

Engineering Resources: Water Systems and Agriculture

The ancient Mesoamerican civilizations possessed an impressive understanding of engineering principles that facilitated the development of intricate water systems and innovative agricultural techniques. Through a sophisticated grasp of hydraulics and environmental adaptation, these societies efficiently managed water resources to support their agrarian needs. The ingenuity displayed in harnessing and directing water sources played a pivotal role in sustaining their complex agricultural endeavors.

One of the engineering marvels employed by Mesoamerican cultures was the construction of elaborate irrigation networks. These systems effectively channeled water from rivers, lakes, and natural springs to irrigate their fields, ensuring consistent crop yields in diverse climatic conditions. The strategic placement of reservoirs, canals, and terraces optimized water distribution across varying terrain, contributing to the abundance of staple crops such as maize, beans, and squash. The cultivation of these vital food staples not only sustained the population but also fostered surplus production, enabling societal growth and prosperity.

Moreover, the technological innovation exhibited in addressing water management challenges extended beyond agricultural sustainability. Mesoamerican engineers devised methods for flood control and soil preservation, mitigating the risks posed by seasonal inundations and promoting long-term land fertility. The efficient utilization of water resources through sustainable practices underscored the symbiotic relationship between engineering and ecological consciousness within these ancient societies.

In addition to pioneering hydraulic engineering, Mesoamerican civilizations demonstrated remarkable expertise in developing agricultural implements and cultivating specialized crops. This included the introduction of innovative tools such as digging sticks, planting sticks, and stone hoes, which significantly enhanced farming productivity and efficiency. Furthermore, the cultivation of crops like cacao, avocados, and various fruits reflected their botanical knowledge and genetic manipulation techniques, paving the way for diverse and resilient agricultural landscapes. The intricate understanding of agroecological systems showcased the Mesoamerican commitment to scientific inquiry and resourceful agricultural practices.

The legacy of Mesoamerican engineering resources and agricultural innovations persists as a testament to the profound impact of indigenous wisdom on sustainable land use and resource management. By integrating advanced engineering solutions with the principles of ecological harmony, these ancient civilizations established enduring models for responsible resource utilization, setting a precedent for contemporary conservation efforts and sustainable agricultural development.

Artistic Technologies: Pottery, Sculpture, and Murals

The artistic achievements of Mesoamerican civilizations are a testament to their sophisticated creative expression and mastery of various artisanal technologies. Central to their cultural identity, pottery, sculpture, and murals served as integral components of Mesoamerican artistry, offering profound insights into the societal structure, religious beliefs, and aesthetic sensibilities of these ancient peoples.

Pottery production in Mesoamerica was a multifaceted process that involved advanced techniques and remarkable attention to detail. Artisans utilized local clay, meticulously shaping and decorating vessels with intricate designs, often incorporating symbolic representations of deities, mythological narratives, and everyday life. The use of specialized tools and meticulous firing methods resulted in durable and visually stunning ceramics, serving both utilitarian and ceremonial purposes within Mesoamerican society.

Sculpture also held a significant role in Mesoamerican art, showcasing the technical prowess and artistic innovation of the region's craftsmen. Works of stone, wood, and other materials were intricately carved and polished to depict divine figures, revered rulers, and mythological creatures. Notably, the Olmec colossal heads stand as remarkable examples of monumental sculptural expertise, underscoring the magnitude of Mesoamerican artistic capabilities and the reverence for iconic representations.

Furthermore, the vibrant murals found in Mesoamerican architectural sites provided vivid visual narratives that communicated cultural, historical, and religious themes. These elaborate wall paintings showcased a rich array of colors achieved through natural pigments, depicting intricate scenes that conveyed narratives of creation mythology, ritual practices, and cosmological concepts. The exquisite preservation of these murals offers invaluable insights into the aesthetics and symbolic meanings embedded in Mesoamerican visual arts, exemplifying the intersection of technology and creativity.

The fusion of artisanal skills, cultural symbolism, and technological ingenuity in pottery, sculpture, and murals underscores the profound complexity of Mesoamerican artistic traditions. These artistic technologies not only served as essential components of daily life and spiritual ceremonies but also stand as enduring testaments to the intellectual and imaginative prowess of Mesoamerican civilizations, capturing the essence of their cultural legacy and enduring impact on the artistic heritage of humanity.

Mathematical Innovations and Their Applications

The mathematical prowess of the Mesoamerican civilizations is a testament to their advanced understanding and application of numerical concepts. The intricate system of counting, based on vigesimal (base-20) math, surpassed the arithmetic capabilities of other contemporary societies. Utilizing a combination of dots and bars as symbols for counting, the Mesoamericans were able to perform complex calculations and create accurate astronomical and calendrical systems. Their sophisticated understanding of geometry enabled the precise construction of architectural wonders such as pyramids and observatories. Additionally, the development of the concept of zero and its incorporation into their numerical notation represented a groundbreaking advancement in mathematical thought.

Moreover, these mathematical innovations were not confined to theoretical knowledge but found practical applications in various aspects of daily life. Agricultural planning and land management benefited from the precise measurement techniques and geometric principles, leading to improved crop yields and sustainable farming practices. Furthermore, the creation of intricate pottery and artistic designs demonstrated the application of mathematical concepts in aesthetic pursuits. The harmony and symmetry evident in Mesoamerican art were achieved through meticulous calculations and geometric arrangements, showcasing the seamless integration of mathematics with creative endeavors.

The accurate timekeeping and calendar systems developed by Mesoamerican mathematicians played a crucial role in religious ceremonies, agricultural rituals, and governance. The ability to predict celestial events and align them with religious festivals was instrumental in maintaining social cohesion and reinforcing the authority of the ruling elite. The mathematical sophistication of the Mesoamerican civilizations also extended to their economic activities, facilitating efficient trade and resource management through numerical record-keeping and advanced calculation methods.

Furthermore, the symbolic significance of numerical representations within the Mesoamerican belief systems transcended mere mathematical utility. Numbers were imbued with sacred meaning and spiritual significance, reflecting a holistic integration of mathematics with cosmological and metaphysical frameworks. This profound intertwining of mathematical innovations with cultural and societal practices underscored the pervasive influence of numerical concepts in shaping the ideological landscape of Mesoamerican civilization.

In conclusion, the mathematical innovations of Mesoamerica represent a remarkable intellectual achievement that permeated every facet of their society. From practical applications in agriculture and art to profound implications for religious and cosmic understanding, the mathematical prowess of the Mesoamerican civilizations stands as a testament to their enduring legacy of intellectual innovation and holistic integration of knowledge.

Interactions with Neighboring Civilizations

Mesoamerica, with its rich cultural tapestry and advanced technological achievements, did not exist in isolation. Interactions and exchanges with neighboring civilizations played a pivotal role in shaping the region's development and technological advancements. These interactions were characterized by the exchange of goods, ideas, and knowledge, contributing to a dynamic environment of innovation and progress.

The Mesoamerican civilizations maintained extensive trade networks with neighboring societies, facilitating the flow of raw materials, luxury goods, and specialized crafts. The exchange of resources such as obsidian, jade, and cacao not only fueled economic growth but also fostered cultural interchange, influencing artistic expressions and material culture in profound ways. This interconnected web of trade routes knit together diverse societies, creating opportunities for collaboration and mutual learning.

Beyond commerce, interactions with neighboring civilizations facilitated the diffusion of knowledge and expertise across geographic boundaries. Mesoamerican mathematicians and astronomers engaged in exchanges with their counterparts in surrounding regions, leading to the assimilation of new mathematical concepts, astronomical techniques, and calendrical systems. This cross-pollination of intellectual pursuits catalyzed further advancements in the fields of mathematics, astronomy, and timekeeping, enriching the collective knowledge base of Mesoamerican societies.

Moreover, the cultural and religious interchange between Mesoamerican civilizations and neighboring societies sparked syncretic developments, resulting in the amalgamation of belief systems, ritual practices, and mythological narratives. These interactions allowed for the harmonization of diverse spiritual traditions, giving rise to complex belief systems that integrated elements from multiple cultures. The fusion of religious ideologies contributed to the evolution of ceremonial practices, architectural styles, and iconography, illustrating the transformative influence of cross-cultural contact on the spiritual and artistic realms.

At the same time, interactions with neighboring civilizations engendered geopolitical dynamics that shaped Mesoamerican sociopolitical landscapes. Diplomatic engagements, alliances, and conflicts influenced the distribution of power and territorial control, leaving indelible imprints on the course of regional history. The interplay of political forces and military strategies reshaped borders, forged alliances, and occasionally led to the dissemination of technologies and innovations across disparate domains, altering the trajectory of societal development.

The enduring impact of these interactions is evident in the enduring legacy of Mesoamerican civilizations, which continues to resonate in contemporary society. The transmission of knowledge and cultural practices across geographical frontiers and the symbiotic relationships cultivated through intercultural exchanges exemplify the profound and lasting effects of Mesoamerican civilizations' interactions with neighboring cultures.

Legacy and Influence on Modern Technology

The legacy of Mesoamerican technology extends far beyond the ancient era, leaving an indelible mark on modern technological advancements and cultural influences. The sophisticated achievements of Mesoamerican civilizations continue to inspire contemporary practices across various fields, from architecture and engineering to mathematics and agriculture.

One of the most prominent aspects of Mesoamerican influence on modern technology lies in the realm of agricultural practices. The innovative techniques developed by these ancient societies, such as terracing, crop rotation, and the cultivation of indigenous plants like maize and chilies, have significantly impacted modern agricultural methods. Their understanding of environmental sustainability and natural resource management serves as a timeless model for contemporary approaches to conservation and ecological balance.

Furthermore, Mesoamerican architectural marvels, including iconic pyramids and intricate city layouts, have influenced modern urban planning and construction techniques. The precise alignment of structures with celestial events also resonates in modern architectural designs and considerations of spatial relationships. In addition, the symbolic significance assigned to architectural forms in Mesoamerican cultures has inspired modern architects to integrate cultural narratives into contemporary buildings, fostering a greater sense of historical continuity and communal identity.

The mathematical innovations of Mesoamerican civilizations, such as the concept of zero and the development of sophisticated calendrical systems, continue to inform modern mathematical principles and timekeeping methods. These foundational contributions underpin contemporary scientific inquiries and technological developments, underscoring the enduring impact of ancient Mesoamerican thought on our understanding of numerical concepts and temporal measurements.

Moreover, the artistic technologies mastered by Mesoamerican artisans, including pottery, sculpture, and vibrant murals, continue to inspire creative expressions in modern art and design. The intricate techniques employed in crafting ornate artifacts and the use of symbolism in visual narratives have transcended time, enriching contemporary artistic endeavors and cultural representations.

Beyond the realms of agriculture, architecture, mathematics, and art, the spiritual and ritual elements embedded within Mesoamerican technologies have left a profound impact on modern cultural practices. Concepts of interconnectedness with nature, ritualized observances of cosmic phenomena, and the belief in the cyclical nature of existence have permeated contemporary philosophical discourses, shaping ethical perspectives and environmental consciousness.

In conclusion, the enduring legacy of Mesoamerican technology encompasses a multi-faceted influence on modern society, touching upon diverse facets of human innovation and cultural evolution. By acknowledging and embracing the enduring contributions of ancient Mesoamerican civilizations, we can cultivate a deeper understanding of our collective heritage and draw inspiration from their timeless wisdom to address contemporary challenges and envision a more harmonious future.

10

The Mysteries of Mesoamerica

The Indus Valley Civilization

The Indus Valley Civilization, also known as the Harappan Civilization, flourished in the vast river plains of what is now modern-day Pakistan and northwest India. Spanning from approximately 3300 BCE to 1300 BCE, this ancient civilization represents one of the world's earliest urban cultures, boasting advanced infrastructure and sophisticated planning. The geographical context of the Indus Valley, with its fertile land and strategic location near the confluence of major rivers such as the Indus and Saraswati, provided an ideal setting for the growth and development of this remarkable civilization. Beyond its material achievements, the Indus Valley Civilization holds profound significance in the annals of ancient history. With a meticulous city layout featuring planned streets, advanced drainage systems, and fortified structures, it reflects a high level of societal organization and engineering prowess. Its economic prosperity and extensive trade networks indicate a complex mercantile society with strong ties to other contemporary civilizations. Moreover, the artistic and cultural artifacts unearthed from this region reveal a sophisticated aesthetic sense and a mastery of craftsmanship, underscoring the civilization's rich cultural heritage. The decipherment of the Indus script, though still a point of academic contention, suggests the existence of a literate society, further illuminating the intellectual and communicative advancements of the time. In conclusion, the Indus Valley Civilization stands as a testament to the ingenuity and resourcefulness of early human societies, leaving a lasting legacy that continues to captivate and inspire modern scholars and enthusiasts alike.

Archaeological Insights: Discovering Technological Artifacts

The exploration of the Indus Valley Civilization has revealed a wealth of technological artifacts that have provided unprecedented insights into their advanced society. Archaeologists and researchers have meticulously uncovered an array of objects that shed light on the technological prowess of this ancient civilization. Among the remarkable discoveries are intricately designed pottery, tools crafted with precision, and artifacts displaying sophisticated craftsmanship. These findings not only attest to the ingenuity of the people of the Indus Valley but also hint at the existence of technologies that were remarkably ahead of their time. The artifacts exhibit a level of refinement and technical sophistication that challenge conventional perceptions of ancient societies. Excavations at sites such as Mohenjo-Daro and Harappa have yielded a treasure trove of technological remnants, offering glimpses into the innovative spirit of the Indus Valley people. These archaeological insights evoke a sense of wonder and curiosity, prompting further investigation into the methods and materials utilized in the creation of these advanced artifacts. The meticulous documentation and preservation of these discoveries provide invaluable clues that illuminate the technological landscape of the ancient civilization. Through a combination of traditional excavation techniques and modern technological advancements, archaeologists continue to unravel the mysteries encapsulated in these technological artifacts. Their findings serve as a testament to the enduring legacy of the Indus Valley Civilization and inspire ongoing explorations into the depths of its technological accomplishments. As contemporary scholars delve deeper into these artifacts, they endeavor to piece together a comprehensive understanding of the innovative capabilities and engineering prowess that characterized this ancient society. The significance of these discoveries extends far beyond mere historical curiosity, offering profound insights into the interconnectedness of technology, culture, and societal advancement. This chapter will delve into the various forms of technological artifacts unearthed in the Indus Valley and examine the implications of these discoveries for our understanding of ancient technological innovation.

Conceptualizing Synthetic Minds: Theoretical Foundations

In delving into the concept of synthetic minds in the context of the ancient Indus Valley civilization, we are confronted with a remarkable intersection of technology and philosophy. The theoretical foundations of synthetic minds invite us to reconsider our understanding of artificial intelligence within this historical framework. At the heart of this exploration lies the fundamental question of what constitutes a 'mind' and how it can be replicated or simulated through technological means. The Indus Valley civilization presents an intriguing case study for contemplating the early manifestations of synthetic thought.

The conceptualization of synthetic minds in this context demands a multidisciplinary approach, drawing upon insights from fields such as cognitive science, anthropology, archaeology, and computer science. It compels us to reevaluate the traditional boundaries that define consciousness and cognition, challenging us to discern the boundaries between organic and artificial intelligence.

One of the key theoretical underpinnings pertains to the notion of intentionality - the capacity of mental states to be directed towards objects or states of affairs in the world. How did the inhabitants of the Indus Valley conceptualize and engineer technology possessing intentional attributes? Delving into this inquiry necessitates engaging with philosophical discourse on the nature of intentionality and its relationship to artificial constructs.

Furthermore, exploring the theoretical foundations of synthetic minds prompts an examination of cultural and societal perspectives prevalent in the ancient Indus Valley civilization. It invites us to contemplate the potential influence of religious, mythological, and cosmological beliefs on the conception and creation of synthetic intelligence. How did the metaphysical outlook of the civilization intersect with their technological pursuits, shaping their understanding of synthetic minds?

Moreover, this exploration encourages a critical analysis of the ethical dimensions inherent to the development and utilization of synthetic minds. Disentangling the ethical considerations prevalent in the ancient Indus Valley civilization offers valuable insights into the moral dilemmas surrounding the creation of artificial intelligence in any societal context. We are compelled to confront issues pertaining to autonomy, responsibility, and the treatment of synthetic beings, shedding light on enduring philosophical quandaries that persist in contemporary debates on AI ethics.

In synthesizing these diverse strands of inquiry, we aim to construct a comprehensive framework for comprehending the theoretical foundations of synthetic minds within the ancient Indus Valley civilization. By plumbing the depths of philosophical, anthropological, and technological knowledge, we endeavor to unravel the enigmatic nature of synthetic intelligence in this ancient realm.

Engineering Techniques and Materials Used

The development of synthetic minds within the Indus Valley Civilization was reliant on advanced engineering techniques and the utilization of various materials that showcased remarkable ingenuity. The artisans and craftsmen of this ancient society demonstrated an exceptional level of expertise in metallurgy, ceramics, and other manufacturing processes, which facilitated the creation of intricate mechanisms that underpinned the synthetic minds.

One notable aspect of the engineering techniques employed by the Indus Valley people was their proficiency in the casting and shaping of copper alloys. Archaeological excavations have unearthed an array of finely crafted bronze artifacts, indicating a sophisticated understanding of metallurgical processes. The ability to control the composition of these alloys and mold them into intricate components highlights the advanced knowledge and skill possessed by the ancient metalworkers.

Furthermore, the sophistication of material application is evident in the use of specialized ceramics for insulating and protecting delicate components within the synthetic minds. The precision in crafting ceramic vessels and encasements, coupled with an understanding of thermal properties, enabled the effective safeguarding of vital components from environmental factors and mechanical stress.

In addition to metallurgy and ceramics, the Indus Valley artisans also demonstrated mastery in the production of specialized gears, bearings, and seals, essential elements for the functioning of the synthetic minds. The precision engineering required for these components underscores the civilization's grasp of geometric principles and machining techniques, illustrating a level of technological advancement that was ahead of its time.

Moreover, the implementation of sustainable and resilient materials such as stone and wood further exemplifies the pragmatic approach taken by the ancient engineers. These foundational materials, combined with the intricate metalwork and ceramic craftsmanship, formed the structural framework for the construction of the synthetic minds, ensuring durability and operational integrity over extended periods.

The synthesis of diverse materials and the seamless integration of complex components within the synthetic minds are a testament to the innovative engineering prowess of the Indus Valley Civilization. By combining their technical expertise with a deep understanding of materials, the ancient engineers achieved remarkable feats, laying the groundwork for the functional advancement of their society and leaving an enduring legacy of technological achievement.

Role of Synthetic Minds in Daily Life

In the Indus Valley Civilization, the concept of synthetic minds held a significant role in the daily lives of its inhabitants. These synthetic minds, which encompassed various forms of advanced technologies and mechanized systems, played a crucial part in shaping the societal framework and facilitating routine activities. One of the primary applications of synthetic minds was in the realm of agriculture and irrigation. Automated irrigation systems outfitted with ingenious water management mechanisms enabled efficient cultivation and contributed to the agricultural prosperity of the civilization. Additionally, these synthetic minds were integral in the development of sophisticated urban planning and infrastructure. From hydraulic engineering marvels to automated gateways and granaries, the utilization of synthetic minds optimized resource allocation and logistics within urban centers. Furthermore, the presence of mechanized textile production indicates the integration of automated systems in crafting textiles, showcasing the advancement in manufacturing processes. The envisioned roles of synthetic minds extended beyond practical applications and permeated into the cultural and ceremonial domains. The construction of awe-inspiring structures and ritualistic sites, supervised by automated construction techniques, reflects the harmonious marriage of technology and spirituality. It is evident that the incorporation of synthetic minds was not confined to mere functionality but also contributed to the enrichment of cultural practices and belief systems. Moreover, evidence suggests the existence of automated scriptoriums and mathematical tools, indicating the intersection of synthetic minds with scholarly pursuits. The seamless inclusion of automated systems in intellectual endeavors underscores the holistic integration of technological innovations into various facets of daily life. The pivotal role of synthetic minds in commerce and trade cannot be overlooked. Automated warehouses, navigational aids, and maritime technologies played an instrumental role in fostering economic exchanges, connecting distant regions, and advancing trade networks. This comprehensive engagement of synthetic minds in bolstering economic activities elucidates the multifaceted impact of technological advancements on the civilization's daily life. Through a multidimensional lens, the significance of synthetic minds becomes increasingly pronounced, illustrating their pervasive influence on the sociocultural and economic dimensions of the Indus Valley Civilization.

Comparative Analysis: Indus vs. Contemporary Civilizations

The comparison of the technological advancements in the ancient Indus Valley Civilization with contemporary societies offers an intriguing insight into the development of human innovation. The Indus Valley Civilization, also known as the Harappan Civilization, flourished around 3300–1300 BCE, showcasing remarkable achievements in urban planning, trade, and technological sophistication. In contrast, modern civilizations have witnessed exponential growth in technology, connectivity, and scientific knowledge. Analyzing the technological landscape of ancient and contemporary civilizations reveals both parallels and divergences that contribute to our understanding of societal progress.

One of the most striking similarities between the Indus Valley and contemporary civilizations lies in the complexity of urban design and infrastructure. The meticulously planned cities of the ancient civilization, featuring advanced sewage systems and robust architectural layouts, bear resemblance to the organized urban centers of present-day societies. Urban planners in both eras have demonstrated a commitment to efficient resource management and sustainable living environments, albeit employing vastly different tools and methods.

Moreover, the emphasis on trade and commerce represents another common thread connecting the two epochs. The Indus Valley witnessed extensive trade networks and sophisticated craftsmanship, evident from archaeological findings such as standardized weights and measures. Similarly, contemporary societies operate within globalized economies, leveraging cutting-edge technologies for commerce and international exchanges. The evolution of trade practices underscores the enduring human endeavor to facilitate economic prosperity through interconnected trade routes and commercial innovations.

Conversely, notable disparities emerge when examining the scope and scale of technological advancements. While the ancient Indus Valley Civilization boasted remarkable engineering feats, including advanced water management systems and intricate artifacts, contemporary societies have harnessed the power of digital technologies, artificial intelligence, and space exploration. The divergence in technological trajectories reflects the dynamic nature of human ingenuity, adapting to the challenges and opportunities of distinct historical epochs.

In essence, undertaking a comparative analysis of the Indus Valley Civilization and contemporary societies illuminates the multifaceted evolution of human innovation. By juxtaposing ancient achievements with modern breakthroughs, we gain a holistic perspective on the continuum of technological progress. This comprehensive understanding enriches our appreciation of historical legacies while providing valuable insights for shaping the future of technological advancement.

Technological Sophistication and Cultural Influence

The technological sophistication of the Indus Valley Civilization not only reflected their advanced engineering abilities but also had a profound impact on their cultural development and societal organization. The intricate nature of their technological achievements provides valuable insights into the societal structure, economic activities, and cultural practices of this ancient civilization.

One of the key aspects of the cultural influence of technological sophistication is evident in the urban planning and architecture of the Indus Valley cities. The carefully planned layout of the cities, including the systematic construction of streets, drainage systems, and multi-story buildings, suggests a level of organization and central authority that was unique for its time. The precision and consistency in the pattern of city construction indicate a well-defined urban infrastructure, reflecting a sophisticated understanding of town planning and civil engineering.

Furthermore, the presence of advanced irrigation systems and water management technologies highlights the significance of agriculture and trade in Indus Valley society. These technologies not only facilitated efficient agricultural practices but also contributed to the economic prosperity of the civilization. The ability to harness and distribute water resources effectively served as a driving force behind the sustenance and growth of urban centers, fostering a thriving agricultural economy and enabling extensive trade networks with neighboring regions.

In addition to practical applications, the cultural influence of technological sophistication can be observed in the artifacts and material culture of the Indus Valley Civilization. The precision and intricacy displayed in the crafting of pottery, jewelry, and other artifacts showcase the artistic and aesthetic sensibilities of the people, demonstrating a blend of functional utility and creative expression. This fusion of technology and artistry underscores the cultural sophistication and the emphasis placed on craftsmanship, design, and symbolic representations within the society.

The impact of technological advancement on cultural practices and social hierarchies is also evidenced by the presence of standardized weights, measures, and seals. These standardized systems not only indicate a level of administrative control but also suggest the existence of a complex trading network and the need for uniformity in economic transactions. The widespread use of seals, often adorned with intricate designs and inscriptions, hints at the development of a writing system or symbolic communication, shedding light on the intellectual prowess and societal organization of the civilization.

Overall, the technological sophistication of the Indus Valley Civilization permeated every aspect of their culture, leaving an indelible mark on their architectural marvels, economic activities, artistic endeavors, and social structures. The integration of advanced technologies into the fabric of society not only propelled the civilization towards unprecedented achievements but also shaped their cultural identity, making the Indus Valley a testament to the harmonious coexistence of technological ingenuity and cultural refinement.

Deciphering the Mechanisms

The Indus Valley Civilization has long remained enigmatic, with its technological achievements shrouded in mystery. As we delve deeper into the artifacts and structures left behind by this ancient civilization, the task of deciphering the mechanisms underlying their synthetic minds becomes a compelling challenge. This section aims to unravel the intricate workings of these technological marvels and provide insights into the innovative thinking that propelled the civilization forward.

At the heart of our quest lies a meticulous study of the relics unearthed from Harappa, Mohenjo-Daro, and other sites. The remarkably advanced urban planning, sophisticated drainage systems, and impeccable grid layouts offer tantalizing glimpses into the engineering prowess of the Indus Valley people. By meticulously examining the remnants of this ancient city-planning, archaeologists and researchers have begun to piece together the inner workings of their synthetic minds.

One of the most intriguing aspects is the presence of standardized weights, measures, and brick sizes across the Indus Valley region. This uniformity suggests a centralized authority capable of implementing and enforcing complex systems, triggering questions about governance, trade, and social organization. The technical precision exhibited in their construction techniques, including woodworking, metalworking, and seal-making, further testifies to their mastery of materials and processes.

Additionally, the enigmatic inscriptions found on seals and tablets have piqued scholarly interest for decades. While attempts to decipher the Indus script continue, some scholars propose that these writings might hold the key to understanding the cognitive capabilities of the ancient inhabitants. The comparison of these symbols with modern computational concepts and linguistic patterns opens up new avenues for interdisciplinary research, shedding light on the potential cognitive functions embedded in these ancient records.

Furthermore, recent advancements in spectroscopy, microscopy, and chemical analysis have allowed us to delve deeper into the material composition of the artifacts, providing invaluable data on the alloys, pottery, and other manufactured items. These insights offer crucial clues regarding craftsmanship, resource utilization, and technological innovation, paving the way for a more comprehensive understanding of the mechanisms at play.

As we piece together these multifaceted puzzles, we are laying the groundwork for bridging the yawning chasm between the past and present. Deciphering the mechanisms of the synthetic minds prevalent in the Indus Valley Civilization not only enriches our knowledge of ancient technology but also holds the promise of inspiring future innovations across diverse domains of human endeavor.

Implications for Modern Technology

The study of synthetic minds in the context of the Indus Valley Civilization presents a fascinating array of implications for modern technology. By delving into the sophisticated engineering techniques and conceptual frameworks employed by this ancient civilization, we gain valuable insights that have the potential to shape the future of technology. It is evident that the technological prowess of the Indus Valley people, particularly in the development of synthetic minds, holds significant implications for contemporary innovation. The understanding of how these synthetic minds operated and integrated into daily life provides a framework for reimagining the role of artificial intelligence in modern society. Moreover, the ability to create complex mechanisms with limited resources underscores the ingenuity and resourcefulness of ancient engineers, serving as an inspiring example for present-day technologists facing similar constraints. Additionally, the comparative analysis of the Indus Valley civilization with contemporary societies raises thought-provoking questions about the evolution of technology and its impact on culture. By exploring the cultural influence of technological advancements, we can reflect on the impact of modern technology on our own cultural practices and societal structures. Furthermore, the legacy of the Indus Valley's synthetic minds triggers reflections on ethical considerations surrounding technological advancements. Understanding how these ancient technologies were embedded in a social and cultural context prompts us to consider the ethical implications of integrating advanced technology into our own communities. This exploration can help guide the responsible development and deployment of future technologies, ensuring that they align with ethical and moral principles. The knowledge gleaned from deciphering the mechanisms and implications of synthetic minds in the Indus Valley offers a platform for interdisciplinary collaboration, sparking innovative thinking across diverse fields. Engaging with the technological heritage of ancient civilizations can inspire interdisciplinary research that transcends traditional boundaries and fosters a holistic approach to modern technological challenges. Moreover, it encourages a deeper appreciation for the complexities of technology and the interplay between cultural, social, and technical dimensions. As we journey through the legacy and continuing mysteries of the Indus Valley's technological marvels, we are poised to reimagine the trajectory of modern technology by drawing from the ancient wisdom that has endured through the ages.

Conclusion: Legacy and Continuing Mysteries

As we conclude our exploration of the synthetic minds of the Indus Valley, it is evident that the legacy of this ancient civilization is deeply intertwined with the mysteries that continue to captivate scholars and innovators alike. The technological achievements of the Indus Valley Civilization not only left a mark on their own era but also exerted influence far beyond their geographical boundaries. This enduring legacy has raised numerous questions, inspiring ongoing research and speculation.

One of the most intriguing aspects of the Indus Valley's synthetic minds is their potential connection to modern technology. The advanced engineering techniques, conceptualization of automated systems, and the application of sophisticated materials hint at a level of innovation that challenges our understanding of ancient civilizations. The implications for modern technology are vast, as the Indus Valley's achievements offer valuable insights for contemporary engineering and design. By studying and understanding these ancient technologies, we can uncover alternative approaches and solutions that may enrich modern practices.

However, alongside the legacy of innovation lies the enigma of the Indus script, an undeciphered writing system that continues to baffle researchers. The inability to interpret this script limits our comprehensive understanding of the synthetic minds and other technological facets of the civilization. The persistent efforts to unravel this mystery underscore the enduring fascination with the Indus Valley Civilization and its enigmatic legacy.

The continuing mysteries surrounding synthetic minds and other technological marvels of the Indus Valley serve as a testament to the timeless allure of ancient ingenuity. The unanswered questions propel ongoing investigations, keeping the spirit of curiosity alive in the pursuit of knowledge. As we grapple with these enigmas, we recognize the value of preserving and studying the remnants of ancient civilizations and the pivotal role they play in shaping our present and future.

In conclusion, the synthetic minds of the Indus Valley stand as a testament to the remarkable technological sophistication of an ancient civilization. Their legacy challenges us to question our assumptions about the capabilities of early societies and encourages us to seek inspiration from the past for contemporary technological advancement. The persistent mysteries that surround these achievements remind us of the enduring allure of ancient history and the endless possibilities it presents for discovery and innovation.

11

Automatons and Oracles: The Greek Insight

The Greek Automation

The origins of mechanical devices in ancient Greek society are rooted in the dawn of antiquity, representing a pivotal leap in the advancement of human civilization. The birth of automation can be traced back to the ingenuity and relentless pursuit of knowledge inherent in Greek culture. As early as the 6th century BC, the Greeks began to devise intricate mechanical contraptions that defied convention and testified to their remarkable technical prowess. These automations were not merely novel inventions but spoke volumes about the philosophical, artistic, and scientific aptitude of the ancient Greeks. The automatons, or self-operating machines, were not only a marvel of engineering but also served as tangible manifestations of the deeply ingrained intellectual curiosity of the age. The Greeks harnessed the power of their imagination to breathe life into these machines, imprinting them with an aura of creativity and cultural significance that echoed through the annals of time. These autonomous marvels offered a unique insight into the Greek mindset, transcending the mere practical applications of technology and shedding light on the interconnected realms of mythology, religion, and innovation. Furthermore, the journey of Greek automation is emblematic of the era's transition towards a more abstract understanding of the world, where mechanisms were imbued with philosophical meaning and aesthetic value. The integration of automation into the fabric of Greek society underpinned their ceaseless quest for knowledge and embodied their deeply rooted desire to comprehend the complexities of existence. Through a profound examination of Greek automation, we unravel not just the technical mastery of an ancient civilization but also the foundational principles that shaped their worldview and continue to influence modern thought.

Historical Context of Greek Technology

The historical context of Greek technology is a tapestry woven with the threads of innovation, intellect, and cultural influence. The ancient Greeks have left an indelible mark on human history through their contributions to various fields, including philosophy, mathematics, art, and technology. To understand the significance of Greek technology, it is essential to delve into the broader historical landscape in which these advancements unfolded.

Ancient Greece was a confluence of civilization, encompassing city-states such as Athens, Sparta, Corinth, and Thebes. It was a time of unprecedented intellectual awakening, culminating in what we now refer to as the Classical Age. This period witnessed the rise of great thinkers like Plato, Aristotle, Socrates, and Pythagoras, who not only laid the foundation for Western philosophy but also made groundbreaking strides in science and mathematics.

Greek technological achievements were intrinsically linked to the sociopolitical dynamics of the era. City-states vied for dominance, spurring competition in various domains, including architecture, engineering, and military tactics. The competitive spirit fostered an environment where ingenuity thrived, leading to innovations that continue to inspire awe and fascination centuries later.

Moreover, the geographical layout of ancient Greece played a pivotal role in shaping its technological landscape. With a rugged terrain and numerous islands, the ancient Greeks became adept seafarers and shipbuilders. Their expertise in naval architecture and maritime navigation laid the groundwork for technological advancements in transportation and trade, enriching their society and fostering cultural exchange with neighboring civilizations.

Greek technology was also influenced by the interactions with other contemporaneous cultures, notably the Egyptians, Persians, and Phoenicians. These exchanges facilitated the transfer of knowledge and technological practices, contributing to the cross-pollination of ideas and the evolution of craftsmanship throughout the Mediterranean and beyond.

In examining the historical context of Greek technology, we uncover a narrative of resilience, adaptability, and intellectual curiosity. It was a time when the pursuit of knowledge and the application of that knowledge propelled the course of human civilization forward. The legacy of Greek innovation continues to serve as a testament to the profound impact that ancient technologies have had on the trajectory of human progress.

Archetypes of Greek Automatons

The exploration of Greek automata reveals a rich tapestry of ingenious mechanical devices, some of which were designed for purely practical purposes, while others held a more symbolic or entertainment-oriented significance. One of the most renowned archetypes is the myth of Talos, the giant bronze automaton crafted by Hephaestus, the god of blacksmiths and artisans. In this tale, Talos was assigned to protect Europa in Crete by patrolling the island's shores three times daily. The concept of a colossal, autonomous guardian constructed from metal exemplifies the ancient Greeks' fascination with the fusion of artistry and utility, setting a precedent for subsequent narratives and technological aspirations across numerous cultures. Moreover, the intricate descriptions of Talos' inner workings serve as a compelling testament to the Greeks' conceptual and technical imagination. Another notable archetype is the automated tripods associated with the Oracle of Delphi. These contraptions were instrumental in facilitating divinations by generating responses to supplicants' queries. Their mechanisms, concealed from public view, added an enigmatic mystique to the religious rituals that permeated ancient Greek society. By examining these examples and others, it becomes evident that Greek automatons transcended mere utilitarian function; they embodied the synthesis of engineering prowess, artistic expression, and philosophical inquiry that characterized the Hellenic quest for understanding and emulation of the natural world. Evidently, the legacy of these archetypes reverberates not only through the annals of history but also within contemporary realms of robotics, artificial intelligence, and experiential design, underscoring the enduring impact of ancient Greek automata on humanity's perpetual pursuit of ingenuity.

Oracles: Divination and Mechanization

The concept of oracles in ancient Greece was a fascinating blend of divination and mechanization, representing an intricate intersection of spirituality and technology. Oracles, such as the renowned Oracle of Delphi, played a pivotal role in the religious and political fabric of Greek society. These revered sites were associated with divine communication, where individuals sought guidance, prophecies, and insight into the future.

The Oracle of Delphi, located at the sanctuary of Apollo, was particularly revered for its enigmatic pronouncements that could influence major decisions, from warfare to governance. The process of divination involved the oracle's medium - often a priestess known as the Pythia - entering a trance-like state to convey cryptic messages purportedly inspired by the gods. This ritualistic tradition, shrouded in mysticism and spiritual authority, captivated the populace and dignitaries alike.

The operation of oracles also featured rudimentary forms of mechanization to enhance the theatricality and mystique of these sacred spaces. For instance, the use of chthonic vapors, subterranean chambers, and deliberate acoustic designs created an eerie ambiance, amplifying the aura of transcendent wisdom. Moreover, the architectural layouts of the sanctuaries played a role in enhancing the mystical allure, often with hidden passages, echoing chambers, and ingeniously designed platforms that contributed to the otherworldly ambiance.

These intricately orchestrated elements not only enriched the visitors' experiences but also showcased the ingenious integration of practical ingenuity with spiritual reverence. The melding of such technical craftsmanship with the profound and spiritual purpose of oracles underscored the deep interconnectedness of ancient Greek society's technological advancements with their metaphysical beliefs.

Moreover, the societal influence of oracles propagated a culture of seeking knowledge and foresight, thereby fostering an inquisitive mindset that transcended philosophical domains. While reliance on oracles gradually waned with the emergence of new belief systems and the evolution of thought in subsequent eras, their historical significance as early machinations of divination and mechanization remains a compelling testament to the intersection of technology and spirituality in ancient Greece.

Key Innovators and Philosophers

The era of ancient Greek technological innovation was profoundly shaped by the contributions of key thinkers, inventors, and philosophers. From the renowned polymath Archytas of Tarentum to the celebrated philosopher Plato, these figures played pivotal roles in advancing the frontiers of ancient technology.

One notable figure is Archytas of Tarentum, a scholar revered for his multifaceted accomplishments in mathematics, music theory, and engineering. His inventive spirit led to the creation of mechanical marvels, including the development of early robotics and basic principles of aerodynamics. Archytas' legacy highlights the interdisciplinary nature of Greek technological ingenuity and its fusion with intellectual pursuits.

Another luminary of this era is Plato, whose profound philosophical insights extended to the realm of technology. While Plato is revered for his enduring contributions to philosophy, his dialogues also expound upon the ethical implications of technological progress. By contemplating the consequences of automata and artificial intelligence, Plato delved into the moral and societal dimensions of innovative endeavors, offering timeless reflections that resonate with contemporary debates on technology.

Furthermore, the discoveries of Heron of Alexandria exemplify the practical applications of ancient Greek technology. His treatises on mechanics and pneumatics showcased pioneering advancements in automatons, water clocks, and sophisticated machinery. Heron's ingenious designs embodied the marriage of theoretical knowledge and practical implementation, underscoring the holistic approach to technological development in ancient Greece.

Additionally, the brilliant mind of Philo of Byzantium enriched the landscape of ancient technology through his comprehensive treatises on diverse engineering disciplines. From hydraulic engineering to military inventions, Philo's meticulous writings illuminated not only the theoretical foundations but also the practical methodologies associated with technological innovation.

The profound influence of these innovators and philosophers reverberates through the annals of history, transcending their respective time periods to inspire generations of scholars, engineers, and visionaries. Their enduring legacies continue to illuminate the interconnectedness of intellectual inquiry, ethical considerations, and practical applications in the realms of ancient technology, reinforcing the notion that the spirit of innovation knows no bounds.

Technological Artifacts and Their Functions

The ancient Greeks were pioneers in developing remarkable technological artifacts that not only showcased their ingenious engineering skills but also served practical purposes in their society. These artifacts were a testament to the innovative spirit of the Greek civilization and continue to captivate modern audiences with their complexity and functionality. Among the most renowned artifacts were the Antikythera mechanism, an astoundingly sophisticated analog computer designed to calculate astronomical positions and eclipses with remarkable precision. Its intricate gear system and advanced mechanisms astound historians and engineers alike, offering invaluable insights into ancient Greek knowledge and expertise in mathematics and astronomy. Another notable technological wonder was the automatic vending machine invented by Hero of Alexandria, known as the world's first known vending machine, which dispensed holy water when a coin was inserted. This invention exemplifies the fusion of religious practices with mechanical ingenuity, demonstrating the multifaceted applications of technology in ancient Greece. Furthermore, the design and construction of architectural marvels such as the Temple of Artemis at Ephesus and the Acropolis exemplify the integration of technological innovation with cultural and artistic expression, showcasing the Greeks' ability to combine functionality with aesthetic appeal. The use of innovative materials like concrete, which allowed for monumental structures to be erected, showcased the Greeks' profound understanding of engineering principles and construction techniques. Notably, the technology implemented in the construction of large-scale theaters, like the Theater of Epidaurus, attests to the Greeks' advancements in acoustics and architectural design, creating spaces with unparalleled acoustic properties that continue to inspire architectural and engineering endeavors in the modern era. The practical functions of these technological artifacts extended beyond mere aesthetics, as they directly contributed to enhancing the quality of life, advancing scientific knowledge, and fostering cultural and societal progression. They accurately reflect the profound impact of technological innovation on the ancient Greek civilization, setting a precedent for future generations and leaving an indelible legacy that continues to influence contemporary society.

Interaction between Culture and Innovation

In ancient Greece, the interaction between culture and innovation was significantly profound, shaping the trajectory of technological progress and societal norms. The intertwining of culture and innovation in Greece was characterized by a unique blend of rationality, creativity, and curiosity, resulting in remarkable advancements that left an indelible mark on the ancient world.

The rich cultural fabric of Greece provided fertile ground for innovation, as artistic expression, philosophy, and scientific inquiry were intertwined with technological ingenuity. This interconnectedness fostered an environment where creativity thrived, enabling the development of groundbreaking inventions and pioneering ideas.

Greek culture, with its emphasis on democracy, intellectual discourse, and the pursuit of knowledge, greatly influenced the direction of innovation. The philosophical underpinnings of Greek society, as exemplified by the likes of Socrates, Plato, and Aristotle, encouraged critical thinking and the exploration of new concepts. These ideals, in turn, spurred the development of innovative technologies, as thinkers sought to apply their philosophical insights to practical solutions.

Moreover, the interconnected nature of Greek city-states facilitated the exchange of ideas and technological know-how, leading to cross-cultural pollination and the dissemination of innovative practices. The bustling agora, where merchants, artisans, and intellectuals congregated, served as a melting pot of diverse perspectives and expertise, fuelling a vibrant marketplace of ideas and inventions.

The influence of Greek innovation extended beyond the confines of the Hellenic world, permeating neighboring civilizations and sparking a wave of technological diffusion. The spread of Greek innovations, such as the water mill, advanced metallurgy, and architectural principles, catalyzed transformative changes in the technological landscape of neighboring societies, leaving an enduring imprint on their cultural and technological evolution.

Furthermore, the integration of technological marvels within cultural practices and religious rites underscored the profound link between innovation and belief systems. The awe-inspiring automata and mechanical devices found in sacred precincts and public spaces not only showcased the prowess of Greek craftsmanship but also imbued these cultural artifacts with spiritual significance, blurring the boundaries between the material and the divine.

In essence, the interaction between culture and innovation in ancient Greece was a dynamic interplay of creativity, intellectual discourse, and cross-cultural exchange, laying the foundation for enduring technological legacies and shaping the broader tapestry of human history.

Impact on Neighboring Civilizations

The technological advancements of ancient Greece had a profound impact on neighboring civilizations, fostering a transference of knowledge and innovation across borders. The sophisticated engineering and inventive automata of the Greeks were revered by neighboring societies, fuelling an exchange of ideas and technologies that influenced the trajectory of historical development. Greek innovations in fields such as mathematics, astronomy, and mechanics permeated throughout the surrounding regions, stimulating intellectual growth and prompting a broader awareness of the potential for technological progress.

From the Hellenistic period through to the Roman era, the spread of Greek ideas and inventions catalyzed a renaissance of learning and discovery. The practical applications of Greek technology, including automata, water clocks, and architectural advancements, were eagerly adopted by neighboring civilizations, leading to an elevation in craftsmanship and construction methods. This influx of knowledge also enriched cultural exchanges and diplomatic relations between these societies, fostering a climate of cooperation and mutual respect. The adaptation of Greek innovations contributed to the evolution of societal structures and governance, serving as foundations for urban planning and administrative systems.

Furthermore, the dissemination of Greek ideas sparked a surge in creative expression and artistic achievements across neighboring lands. The integration of mechanical ingenuity into various forms of art, such as sculpture and performance, showcased the seamless amalgamation of science and aesthetics. This fusion not only elevated the quality of artistic endeavors but also widened the scope of human expression, ushering in an era of boundary-defying creativity and innovation, while enriching the cultural tapestry of the ancient world.

As Greek knowledge permeated neighboring civilizations, it also inspired critical thinking and philosophical discourse, fueling a movement of introspection and contemplation. The incorporation of Greek technological principles into practical applications, combined with the exploration of metaphysical concepts, laid the groundwork for profound intellectual and spiritual exploration. Thus, the impact was not bound merely to material achievements, but extended to the transcendence of the intellectual and spiritual horizons of the era.Overall, the influence of Greek technology on neighboring civilizations triggered a transformative period in history, propelling societies toward new frontiers of knowledge, creativity, and interconnectedness. The legacy of this exchange continues to resonate in the fabric of human achievement, reminding us of the enduring power of innovation and the unifying force of shared knowledge.

The transition from Greek to Roman Technological Endeavors

With the gradual decline of Ancient Greece, the Roman Republic emerged as a dominant force in the Mediterranean world. This transition also brought about significant shifts in technological and engineering advances. While the Romans were highly influenced by Greek knowledge and craftsmanship, they expanded on these foundations, creating their own legacy of innovation and engineering prowess. One of the key areas of development lay in the field of civil engineering, with the construction of vast aqueducts, roads, and monumental structures such as the Colosseum and the Pantheon. The Romans perfected the use of arches and domes, creating structures that still stand as marvels of ancient engineering today. Additionally, the Roman mastery of concrete allowed for the construction of enduring infrastructure that revolutionized urban planning and architecture. The Roman military also played a crucial role in driving technological progress. Their sophisticated siege engines, including ballistae and catapults, demonstrated a deep understanding of mechanics and materials. Furthermore, Roman advancements in metallurgy led to the production of superior weaponry and armor. The utilization of standardized measurement systems and the implementation of innovative tactics further underscored their technological achievements. In the realm of medicine, the Romans made significant strides in public health and sanitation, enhancing the quality of life for their citizens. They developed advanced aqueducts and sewage systems to provide clean water and efficient waste disposal, setting new standards for urban hygiene. Building upon the foundations laid by the Greeks, the Romans integrated diverse influences from conquered lands, assimilating ideas and refining techniques to forge an expansive technological empire. Their ability to adapt and enhance existing knowledge paved the way for the spread of their innovations across vast territories, leaving a lasting imprint on subsequent civilizations. As the Roman Empire flourished, its technological prowess became a defining characteristic, shaping the course of history and influencing generations to come. This dynamic transition from Greek to Roman technological endeavors heralded an era of remarkable ingenuity and advancement, laying the groundwork for future technological revolutions and leaving an indelible mark on the annals of human achievement.

Summary

As we bid farewell to the age of Greek innovation, a compelling transition awaits as we step into the realm of Roman technological endeavors. The legacy of Greek technology has left an indelible mark on the fabric of human progress, yet the rise of the Roman Empire signals a new chapter in the annals of history, one characterized by its own unique brand of ingenuity and engineering prowess.

The Romans, inheritors of Greek knowledge and wisdom, extended their technological reach far and wide, exhibiting a penchant for massive infrastructure projects, advanced hydraulic systems, sophisticated construction techniques, and ingenious military inventions. This period witnessed the grandeur of aqueducts, the marvel of complex road networks, and the mastery of architectural feats that continue to inspire awe and admiration to this day.

One cannot simply overlook the significance of Roman advancements in the fields of civil engineering, architecture, and urban planning. The era's empirical approach to problem-solving and utilization of pragmatic solutions solidified the Roman Republic and Empire as unparalleled leaders in the domain of practical innovation, shaping the very foundations of modern civilization.

Moreover, the cultural exchange and assimilation between the Greek and Roman worlds cultivated a fusion of ideas, resulting in a dynamic technological landscape that synthesized both Hellenistic and indigenous Roman traditions. The embodiment of this convergence is evident in the development of state-of-the-art war machines, monumental public works, and the refinement of previously established scientific principles.

It is imperative to recognize the enduring impact of Roman technological achievements, which reverberates across millennia and continues to shape our contemporary world. The spirit of their resilience, ambition, and transformative vision resonates through the ages, offering profound insights into the interplay between innovation and societal progress—a narrative that transcends time and space.

In conclusion, as we pivot from the Greek insight to the dawn of Rome's technological endeavors, we are poised to embark upon a captivating expedition into an epoch defined by unprecedented accomplishments, enduring legacies, and the perpetuation of human ingenuity at its zenith.

12

The Mechanical Wonders of Ancient Rome

Roman Engineering

Roman engineering stands as a testament to the ingenuity and resourcefulness of an ancient civilization. It roots extend deep into history, reflecting the foundations and progressions of engineering that have shaped the world we know today. From its earliest phases, Roman engineering showcased an unprecedented level of sophistication, innovation, and practicality that allowed the empire to flourish and leave an indelible mark on global infrastructure. The mastery of diverse construction techniques, including arches, domes, and concrete, facilitated the creation of monumental structures such as aqueducts, amphitheaters, and baths. These achievements demonstrated the Romans' remarkable grasp of structural mechanics, materials science, and urban planning, all of which continue to inspire engineers and architects across centuries. The evolution of Roman engineering paralleled the expansion and consolidation of the empire, as ambitious projects aimed to connect and solidify the far reaches of the realm. Through this exploration, we gain insight into the convergence of technical prowess, strategic vision, and sheer perseverance that propelled Roman engineering to unparalleled heights. By unraveling the intricacies of their methods and accomplishments, we not only honor the legacy of Roman engineers but also enrich our understanding of the sophisticated solutions they devised to overcome the challenges of their time.

Architectural Mastery: Building the Empire

The architectural achievements of ancient Rome stand as a testament to the ingenuity and vision of Roman engineers and builders. The grandeur of Roman architecture, from colossal amphitheaters to majestic temples and awe-inspiring aqueducts, symbolizes the ambitions and capabilities of an empire at its peak. Roman architectural mastery was not only about creating imposing structures but also ensuring their durability and functionality for generations to come. The intricate designs and methods employed by Roman architects and engineers have left an indelible mark on the world.

One of the most remarkable feats of Roman architectural mastery is epitomized in the construction of the iconic Colosseum. This mammoth structure, with its intricate network of arches and tiers, not only served as a venue for gladiatorial contests and public spectacles but also stood as a statement of Rome's prowess in engineering and design. The innovative techniques used in the construction of the Colosseum, including the use of concrete and advanced scaffolding systems, revolutionized the field of architecture and set a new standard for monumental building projects.

In addition to the Colosseum, Rome's Pantheon stands as a timeless testament to the architectural brilliance of the ancient Romans. The engineering marvel of the Pantheon's dome, with its perfect hemisphere and oculus, showcases the Romans' command over material science and structural integrity. The meticulous symmetry and harmonious proportions of the Pantheon continue to inspire awe and admiration, serving as a living example of Rome's architectural finesse.

Furthermore, the Roman Forum emerges as the heart of ancient Rome's urban development and civic life. The meticulously planned layout of temples, basilicas, and government buildings, intertwined with open squares and bustling marketplaces, exemplifies the Romans' grasp of urban planning and architectural cohesion. The strategic positioning of these public spaces within the cityscape reflected the Romans' understanding of social dynamics and the importance of communal interaction.

The architectural advancements made by the Romans also extended beyond grand monuments to encompass everyday infrastructure. The construction of durable bridges, innovative road networks, and expansive public baths underscored the Romans' commitment to improving the lives and livelihoods of their citizens. The legacy of Roman architectural mastery continues to influence contemporary construction and urban planning, serving as a source of inspiration for architects and engineers around the world.

Aqueducts and Water Management Systems

The engineering marvels of Ancient Rome extended beyond architectural feats and military innovations to encompass sophisticated aqueducts and water management systems that played a pivotal role in sustaining the vast empire. Aqueducts represented an unparalleled achievement in hydraulic engineering, enabling the transportation of fresh water from distant sources to urban centers and remote outposts. These monumental structures, constructed with meticulous precision and ingenuity, underscored Rome's commitment to harnessing natural resources for the benefit of its populace.

The complex network of aqueducts incorporated various innovative techniques, such as gravity-fed water channels, siphons, inverted siphons, and arcades, to maintain a consistent flow of water across diverse terrains. The mastery of hydraulic principles and surveying allowed Roman engineers to navigate challenging landscapes, including mountains and valleys, while ensuring efficient water distribution to support burgeoning populations. The strategic positioning of reservoirs and water towers facilitated the regulation and storage of water, safeguarding against scarcity and ensuring a reliable water supply for public baths, fountains, and private residences.

Moreover, the technological sophistication of Roman aqueducts encompassed advanced filtration and purification methods to deliver potable water free from contaminants. The use of settling tanks, sand filters, and conduits lined with waterproof materials exemplified the comprehensive approach toward maintaining water quality, reflecting an awareness of public health and sanitation unparalleled in antiquity. These pioneering endeavors in water management not only elevated the standard of living for Roman citizens but also set a precedent for urban planning and infrastructure development that resonates with modern society.

Furthermore, the enduring legacy of Roman aqueducts echoes through the longevity and resilience of these engineering marvels. Many aqueducts continued to function for centuries, serving as a testament to the durability and foresight inherent in their construction. The monumental scale and engineering prowess demonstrated in the creation of aqueducts stand as enduring testaments to Rome's unparalleled ingenuity and technical proficiency, leaving an indelible mark on the evolution of hydraulic engineering and infrastructure practices across civilizations.

Military Innovations and Siege Machinery

The military innovations and siege machinery of ancient Rome represent an extraordinary testament to the ingenuity and engineering skills of the Roman civilization. Warfare was an integral part of the ancient world, and the Romans were pioneers in developing advanced weaponry and tactics to conquer and defend their vast empire.

One of the most iconic and awe-inspiring aspects of Roman military engineering was the development of formidable siege machinery. The Romans were renowned for their formidable catapults, ballistae, and siege towers, which were used to launch projectiles over high walls and fortifications, allowing them to breach the defenses of enemy cities and fortresses. These sophisticated war machines were meticulously engineered and constructed, demonstrating the Romans' remarkable understanding of mechanics and physics.

The meticulous planning and construction of siege engines such as the onager and scorpion showcased the Romans' mastery of projectile technology. These powerful machines could hurl massive stones, incendiary projectiles, and even diseased animal carcasses with devastating accuracy, wreaking havoc upon enemy strongholds and demoralizing their defenders.

In addition to siege machinery, the Romans excelled in the development of innovative military technologies. The construction of roads and bridges facilitated the rapid movement of troops and supplies, enabling the Roman legions to exert strategic dominance over vast territories. Furthermore, the Roman army employed sophisticated defensive structures such as ramparts, entrenchments, and palisades to fortify their positions during sieges and military campaigns.

The ingenuity of Roman military engineers extended beyond traditional warfare and into the realm of naval architecture. The construction of powerful warships equipped with advanced ballistae and boarding mechanisms allowed the Roman navy to assert maritime supremacy and project Roman power across the Mediterranean and beyond.

The legacy of Roman military innovation continues to influence modern military engineering and strategy. The principles of siege warfare developed by the Romans have endured through the centuries, shaping the evolution of military technology and tactics. The enduring impact of Roman military innovation serves as a testament to the enduring legacy of Rome's technological prowess and its profound influence on the course of human history.

Public Works and Urban Infrastructure

The ancient Romans were renowned for their remarkable feats of engineering and their mastery of urban infrastructure. The city of Rome itself served as a testament to their ingenuity, boasting an impressive array of public works and urban infrastructure that were unrivaled in the ancient world. Central to Roman urban planning was the concept of the 'Forum', a bustling hub of political, commercial, and social activity. The Forum Romanum, in particular, was the heart of ancient Rome, housing various important buildings, temples, and civic spaces. Its architectural grandeur exemplified the Romans' commitment to creating vibrant and functional urban centers. Beyond the iconic Forum, the Romans excelled in the construction of public baths, amphitheaters, and marketplaces, all of which contributed to the city's vitality and cultural vibrancy. The construction of public baths, known as thermae, was a significant aspect of Roman urban life. These expansive complexes not only provided a space for personal hygiene but also served as centers for socializing, exercise, and relaxation. The Baths of Caracalla, one of the largest public baths in Rome, reflected the luxurious and communal nature of these establishments. Similarly, the Colosseum, with its seating capacity of over 50,000 spectators, stood as a testament to Rome's engineering prowess and dedication to entertainment. Beyond the capital, Roman cities across the empire showcased an intricate network of roads, bridges, sewer systems, and aqueducts, forming the backbone of urban infrastructure. The construction of roads, such as the famous Appian Way, facilitated efficient communication and transportation, connecting the farthest reaches of the empire to the city of Rome. Aqueducts, engineered to transport water over long distances, ensured a stable water supply for public fountains, baths, and private residences. The Pont du Gard in France and the Aqueduct of Segovia in Spain are enduring examples of Roman aqueduct design and construction. In addition to their innovation in public works, the Romans implemented advanced urban planning and sanitation practices. They developed complex sewer systems, known as Cloacae, to manage the city's waste and runoff efficiently. Civic buildings and public spaces were strategically located to optimize crowd control, commerce, and civic activities. As a result, the Romans created a system of urban infrastructure that not only supported the physical needs of the populace but also enriched their cultural and social experiences.

The Great Roman Roads: Connectivity and Mobility

The road network constructed by the ancient Romans stands as an enduring testament to their unparalleled engineering prowess. Renowned for their remarkable sense of urban planning and infrastructure, the Romans developed a sophisticated system of roads that formed the backbone of their vast empire. Stretching over 250,000 miles across Europe, Asia Minor, and North Africa, these roads played a pivotal role in establishing unrivaled connectivity and facilitating efficient mobility throughout the empire.

Characterized by their engineering brilliance, Roman roads were meticulously designed and constructed using a layering technique. The foundation consisted of large stones or gravel, followed by a layer of finer gravel, topped with carefully fitted paving stones. This meticulous construction method ensured durability and longevity, enabling the roads to withstand heavy traffic and harsh weather conditions. Furthermore, the roads featured cambered surfaces and well-drained foundations, allowing rainwater to be swiftly diverted, thereby minimizing erosion and maintaining the structural integrity of the thoroughfares.

Beyond their remarkable construction, Roman roads revolutionized travel and transportation during ancient times. These arteries of the empire facilitated the swift movement of troops, enabling the rapid deployment of legions to various frontiers of the Roman dominion. Additionally, the roads supported the efficient transport of goods, bolstering trade and economic prosperity throughout the expansive Roman territories. The strategic positioning of military outposts along these thoroughfares contributed to enhanced security and control, underscoring the crucial role of the road network in consolidating Roman authority.

Moreover, the Roman roads served as conduits for the exchange of ideas, culture, and knowledge, fostering a sense of interconnectedness and unity across diverse regions. The development of vibrant urban centers, thriving market towns, and bustling trade hubs along these routes attests to the transformative impact of the road network on the social and economic fabric of the ancient world. It is evident that the Roman roads not only facilitated physical movement but also facilitated the dissemination of intellectual and cultural exchanges, leaving an indelible mark on the historical tapestry of human civilization.

Even today, remnants of the ancient Roman road network endure as enduring vestiges of an era characterized by ingenuity and foresight. While many of these roads have been repurposed or modernized, their enduring legacy remains a testament to the profound impact of Roman engineering and the enduring influence of their technological innovations on shaping the course of history.

Material Science in Ancient Rome

The innovative prowess of ancient Rome extended beyond engineering and architecture, encompassing a sophisticated understanding of material science. Roman craftsmen and engineers were adept at manipulating a variety of materials to create structures and artifacts that have endured the test of time. One of the most iconic materials associated with Roman construction is concrete, a blend of lime, volcanic ash, and rubble that allowed for the creation of monumental structures such as the Pantheon and the Colosseum. The Romans' mastery of concrete was not only evident in the scale of their architectural wonders but also in its durability, proving resistant to the ravages of time and weather. This enduring legacy continues to inspire modern building techniques and materials science. Additionally, the Romans excelled in metallurgy, producing high-quality steel, iron, and alloys that were vital to their military conquests, infrastructure development, and artistic achievements. The development of precise techniques for mining, smelting, and forging played a pivotal role in sustaining the empire's expansion and the sophistication of its engineering projects. The innovative approaches to material refinement and production contributed significantly to the success and longevity of Roman civilization. Furthermore, the use of advanced materials in pottery and glassmaking showcased the Romans' ingenuity and aesthetics. Artisans perfected the creation of exquisite ceramics and glassware, utilizing techniques such as mold casting and glass blowing to achieve remarkable levels of sophistication and artistry. These products not only served functional purposes but also reflected the wealth and cultural refinement of Roman society. The utilization of materials in ancient Rome went beyond mere utility; it embodied a profound appreciation for craftsmanship, innovation, and timeless creativity. The legacy of Roman material science echoes through the annals of history, impacting subsequent generations and serving as a testament to the enduring capabilities of human ingenuity.

Automation and Labor-saving Devices

The ancient Romans were adept at devising ingenious automation and labor-saving devices that significantly contributed to the efficiency and productivity of various industries. Among the most notable advancements was the use of water-powered systems known as 'hydraulics.' This technology enabled the Romans to create intricate and efficient mechanisms for tasks such as milling grains, sawing timber, and powering heavy machinery in mining operations and construction sites. Water wheels, along with advanced aqueduct systems, were harnessed to provide power for a range of industrial processes, revolutionizing the Roman economy and setting new standards for technological progress.

In addition to hydraulics, the Romans also developed an array of automatic and semi-automatic tools and devices to streamline everyday tasks. Examples include automatic doors in public buildings that operated using a system of counterweights, self-regulating water clocks that facilitated accurate timekeeping, and complex automated looms that enhanced textile production. These innovations not only showcased the Romans' engineering prowess but also offered practical solutions to alleviate human labor.

One particularly remarkable invention was the 'treadwheel,' a human-powered device that functioned as a labor-saving mechanism for lifting heavy weights. Used extensively in construction, agriculture, and various industrial settings, the treadwheel comprised a large rotating wheel with steps or treads affixed around its circumference. Workers would walk inside the wheel, causing it to rotate and lift or move heavy objects, thereby reducing physical strain while increasing efficiency. The widespread adoption of the treadwheel reflected the Romans' commitment to finding innovative solutions to enhance productivity and minimize manual effort.

Furthermore, the concept of 'automata' flourished in ancient Rome, demonstrating the Romans' fascination with mechanized figures and devices capable of autonomous movement. Intricate mechanical figurines, clockwork-driven displays, and even automated theatrical props contributed to the entertainment and technological prestige of Roman society. These early examples of sophisticated automata served as precursors to the development of modern robotics and automation, highlighting the enduring influence of Roman ingenuity on the evolution of technology.

The integration of automation and labor-saving devices into Roman daily life not only elevated industrial capabilities but also paved the way for future innovations, laying the foundation for the Industrial Revolution and subsequent technological revolutions. The legacy of these ancient Roman inventions serves as a testament to the enduring impact of early automation and labor-saving technologies on the progression of human civilization.

Impact on Society and Economy

The technological advancements in ancient Rome had a profound impact on both the society and the economy of the Roman Empire. One of the most significant contributions was the introduction of labor-saving devices and automation in various industries. This not only increased efficiency but also freed up human labor for other tasks, leading to overall economic growth and innovation. The widespread use of automated systems in agriculture, such as water-powered flour mills and automated grain threshing machines, revolutionized food production and distribution. As a result, the Roman Empire was able to sustain a larger population and support the development of urban centers. This shift in agricultural practices fostered an increase in trade and commerce, stimulating economic growth across the empire. Additionally, the implementation of innovative construction techniques and mechanized building methods played a pivotal role in the expansion and modernization of infrastructure, including roads, aqueducts, and public buildings. The efficient transportation networks and distribution channels facilitated by Roman engineering further bolstered trade and economic prosperity. Moreover, the employment of advanced materials and machinery in mining and metalworking enhanced resource extraction and manufacturing processes, directly shaping the economic landscape. Apart from the direct economic implications, the adoption of sophisticated technologies influenced social stratification and the division of labor within Roman society. The availability of automation and labor-saving devices enabled more specialized roles to emerge, contributing to the diversification of labor and skills. Craftsmen, engineers, and technicians became esteemed members of society, driving further technical innovation while transforming the socio-economic structure. Furthermore, the widespread dissemination of advanced technologies across the empire facilitated the exchange of ideas and cultural diffusion, fostering a climate of intellectual collaboration and learning. This cross-pollination of knowledge and expertise created a dynamic environment that catalyzed artistic and scientific achievements. The technological prowess of ancient Rome not only propelled economic growth and societal organization but also laid the foundation for the development and spread of innovative practices that continue to influence modern technology and industry. Elements of Roman engineering and innovation persist in contemporary infrastructure, architecture, and manufacturing techniques, underscoring the enduring legacy of ancient Roman technology on the modern world's social and economic dynamics.

Legacy and Influence on Modern Technology

The legacy of Ancient Rome in shaping modern technology is indisputable, with enduring influence across a multitude of domains. The engineering feats and advanced technological innovations of the Roman Empire have continued to inspire and inform present-day engineering practices, architectural designs, and material sciences. This profound legacy has significantly impacted modern infrastructure, urban planning, military technology, and even social structures. One notable area where the influence of ancient Roman technology is prominently evident in contemporary times is in the construction and design of robust, enduring infrastructure. The Romans' sophisticated aqueduct systems, road networks, and architectural marvels have set a precedent for durable infrastructure, serving as a model for modern civil engineering projects worldwide. Moreover, the utilization of concrete in Roman architecture, an innovation developed over two millennia ago, continues to inspire modern construction materials and techniques. The durability and resilience of Roman concrete have been studied and emulated in the development of sustainable, long-lasting building materials. Additionally, the monumental Roman road network, renowned for its efficiency and durability, laid the foundation for contemporary transportation and connectivity systems, impacting the modern design and construction of highways, interstates, and transportation infrastructure. Furthermore, the military technologies and innovations pioneered by the Romans have left a lasting imprint on modern defense strategies and warfare tactics. The engineering prowess displayed in the development of siege machinery and military fortifications has influenced the evolution of contemporary weaponry and defensive structures. The strategic principles and mechanized devices employed by the Roman military continue to provide valuable insights for modern defense systems and military engineering. Beyond technical innovations, the societal and economic implications of Roman technology endure through numerous cultural, legal, and governance influences. The Roman approach to urban planning and public works has informed the fundamental concepts of city development and infrastructure management in contemporary society. Additionally, the Roman model of governance, administrative systems, and legal frameworks has contributed to the establishment of modern governance structures, influencing legal and political systems worldwide. The enduring legacy of the Roman Empire's technological achievements resonates not only in tangible, physical artifacts but also in the intangible influences that permeate various aspects of modernity, including culture, governance, and civilization. This rich heritage continues to inspire and inform the trajectory of technological progress, affording a deeper understanding of our collective historical roots while offering invaluable insights for future innovation and advancement.

13

The Legacy of Atlantis

Introduction: Myth or Reality?

The enigmatic legend of Atlantis has captivated the human imagination for centuries, presenting an intriguing dichotomy between myth and reality. Predominantly depicted as a utopian civilization submerged beneath the depths of the ocean, Atlantis epitomizes the enduring mystery that shrouds ancient history. Despite persistent efforts by historians, archaeologists, and scholars to authenticate its existence, the lack of empirical evidence has perpetuated an ongoing debate concerning the veracity of its legacy. The story of Atlantis embodies a paradoxical coexistence in the collective consciousness—simultaneously dismissed as fantasy and revered as historical truth, casting a pervasive allure across various cultures worldwide. Numerous interpretations of this legendary city span the realms of literature, philosophy, and art, propelling its influence far beyond traditional geopolitical boundaries. The ambiguity surrounding Atlantis' authenticity reveals a profound tension between skepticism and curiosity within the scholarly community and general populations alike. Although concrete proof remains elusive, the resonance of the Atlantis narrative maintains a firm grip on popular imagination, serving as a testament to the enduring human fascination with antiquity and the unknown. As we delve into the intricate tapestry of historical accounts and unearthed descriptions, it becomes imperative to navigate through this labyrinthine intersection of mythos and tangible evidence, to unravel the enigma of Atlantis and discern whether its legacy is rooted in the annals of actuality or the boundless realm of conjecture.

Historical Accounts and Descriptions

The historical accounts of Atlantis have been a subject of debate for centuries, captivating the imagination of scholars, explorers, and enthusiasts alike. While some regard Atlantis as a mere myth shrouded in allegory, others tirelessly seek evidence to validate its existence. The earliest mention of Atlantis can be traced back to the works of the ancient Greek philosopher Plato, who vividly described this enigmatic island in his dialogues 'Timaeus' and 'Critias'. According to Plato's accounts, Atlantis was an advanced civilization that mysteriously vanished in a catastrophic event. The detailed descriptions portray Atlantis as a prosperous society with remarkable technological achievements and unparalleled wealth, making it a source of fascination and speculation throughout history. Several historical texts and archaeological findings from diverse cultures and periods have also alluded to the possible existence of Atlantis, generating a rich tapestry of anecdotes, theories, and inquiries that continue to intrigue researchers and enthusiasts worldwide. While skeptics assert that these narratives are merely part of a collective lore, proponents argue that there are striking parallels between the depictions of Atlantis and remnants of advanced engineering and cultural influences across different traditions. As historians and archaeologists delve deeper into deciphering the enigma of Atlantis, they aim to discern the veracity of these accounts from a multidisciplinary perspective, drawing upon fields such as geology, anthropology, and oceanography to unravel the historical truth behind this legendary city.

Architectural Innovations and Descriptions

The enigmatic legend of Atlantis has long captivated the imaginations of historians, scholars, and enthusiasts alike. As we delve deeper into the architectural innovations and descriptions attributed to this mythical civilization, we are faced with a plethora of speculative narratives intertwined with the realities of ancient cultural advancements. The purported grandeur and sophistication of Atlantean architecture has been a subject of fervent debate and scrutiny, often blurring the lines between myth and historical actuality.

Scholars have proposed that Atlantean structures were unparalleled in their scale and precision, featuring engineering marvels that surpassed the capabilities of contemporary civilizations. From towering citadels adorned with intricate carvings to expansive subterranean chambers intricately aligned with celestial phenomena, the architectural legacy of Atlantis is shrouded in mystery, leaving us to ponder the extent of their technical expertise and innovative prowess.

Key focal points encompass the utilization of advanced geomancy in urban planning, the integration of sustainable materials within monumental constructions, and the alleged harnessing of energy sources that defy conventional understanding. Emblematic of this advanced architectural repertoire is the emphasis on harmonizing structural elements with the natural environment, illustrating a profound reverence for the interconnectedness of natural forces and human ingenuity.

Descriptions of ornate edifices adorned with ornamental motifs serving both aesthetic and functional purposes persist in various ancient texts and oral traditions. Encoded within these depictions are hints of hybrid forms and intricate geometries, suggestive of a synthesis between artistic expression and technological innovation. Moreover, speculation surrounding the existence of acoustic resonance chambers and water management systems reflecting an adept understanding of fluid dynamics further underscores the enigmatic nature of Atlantean architecture.

Efforts to discern the veracity of these claims through rigorous archaeological exploration and analysis continue to yield intriguing insights, necessitating a nuanced approach towards interpreting the enduring legacy of Atlantis. The compelling juxtaposition of transcendental mythos and empirical evidence propels our quest to unravel the enigma that is Atlantean architecture, invoking contemplation on the potential intersections between speculative narratives and tangible historical artifacts.

Technological Artifacts and Their Functions

The enigmatic legacy of Atlantis beckons us to explore its technological artifacts, shedding light on the advancements that purportedly flourished within its ancient civilization. As we delve into this captivating realm, a myriad of objects come to the forefront, each seemingly attesting to the sophistication of Atlantean ingenuity.

Among the most striking artifacts are the intricate energy crystals, believed to have powered the city while harnessing natural forces for a variety of purposes. These crystals, adorned with geometric patterns and inscriptions denoting their supposed functions, have incited fervent speculation among scholars and enthusiasts alike. Some theories posit that these crystals facilitated energy distribution throughout the city, powering remarkable feats of engineering and sustaining a harmonious way of life.

In addition to the enigmatic crystals, explorations of purported Atlantean sites have yielded an array of advanced devices and machinery. From intricately crafted mechanical contraptions to complex navigational instruments, these artifacts hint at a society far ahead of its time. One particularly fascinating discovery is what appears to be a weather-manipulating device, designed to control atmospheric conditions in a display of mastery over nature's elements.

Moreover, the uncovering of remarkably preserved textiles and metallic alloys has unveiled the sophistication of Atlantean craftsmanship. Inscriptions and pictorial depictions on various artifacts suggest a profound understanding of astronomical phenomena, potentially indicating celestial navigation techniques and astronomical observations. These findings evoke a sense of awe and prompt reevaluation of our understanding of ancient civilizations, challenging the conventional timeline of technological progress.

While the function and operation of these artifacts remain subjects of speculative conjecture, their existence raises thought-provoking questions about the technological prowess and societal dynamics of the fabled island civilization. The significance of these artifacts extends beyond mere historical curiosity, beckoning us to reassess the boundaries of human achievement and the spectrum of possible knowledge lost to the depths of time.

Atlantean Influence on Surrounding Cultures

The enduring allure of Atlantis lies not only in the mystery surrounding its existence, but also in the supposed influence it allegedly exerted on surrounding ancient cultures. Legends and accounts from various civilizations widely speak of a highly advanced civilization that thrived in an era lost to antiquity, and many of these stories contain intriguing similarities, suggesting a diffusion of Atlantean knowledge and technology. The profound impact attributed to Atlantis manifests in diverse cultural and technological aspects across different regions. These influences range from architectural styles and engineering techniques to mythical motifs and spiritual beliefs, reflecting a potential web of connectivity that radiated outward from the fabled island. Echoes of this purported influence can be found in the design and construction of megalithic structures, such as the enigmatic alignments of the pyramids in Egypt and the remarkable precision of the Andean monuments. Moreover, the dissemination of maritime knowledge and navigational skills across ancient seafaring cultures has been conjectured to bear the indelible imprint of Atlantean expertise in cartography and ocean exploration. The infusion of technological wisdom, if indeed traced back to Atlantis, would have instigated a paradigm shift, catalyzing intellectual and cultural advancements in spheres such as architecture, engineering, and astronomy. Scholars perennially debate the extent of this alleged diffusion, seeking corroborative evidence through comparative studies of artifacts, linguistic connections, and shared symbolic representations. The tantalizing notion of an advanced predecessor civilization influencing diverse societies continues to captivate researchers, spurring meticulous examinations of archaeological sites and textual sources. While skeptics dismiss the purported Atlantean influence as mere myth and folklore, proponents posit that the extraordinary unity in certain methodologies, materials, and artistic motifs among seemingly disparate cultures warrants consideration. Furthermore, the enigmatic correspondence between accounts of a cataclysmic submergence and flood myths harkens to a shared narrative that transcended regional boundaries, hinting at a profound cultural exchange facilitated by Atlantean agency. Unraveling the veracity of these claims remains an ardently pursued quest, driving interdisciplinary investigations that endeavor to elucidate the scope and significance of any potential Atlantean impact on the development of ancient civilizations.

Engineering Marvels: Between Myth and Reality

The Atlantean civilization has long been shrouded in mystery, with the myth of its existence often eclipsing any attempts at historical analysis. However, within this enigmatic aura lies a trove of intriguing engineering marvels that have fueled countless debates between myth and reality. The legacy of Atlantis continues to captivate scholars and enthusiasts, beckoning them to unravel the truth behind its technological feats. Even as some deem it purely mythical, others propose that its influence extended far beyond mere allegory. One can't help but wonder — could there be some kernels of truth buried within the layers of myth? Among the fables and mystique, whispers of advanced infrastructure and awe-inspiring architecture have persisted through generations. The descriptions of Atlantean structures, including colossal temples, concentric rings of water and earth, and intricate irrigation systems, evoke visions of a society far ahead of its time. Furthermore, tales of ingenious machinery and complex mechanical devices have sparked fervent imaginations, prompting extensive conjecture regarding the extent of Atlantean technological prowess. While concrete evidence may be scant, the enticing prospect of uncovering tangible remnants of Atlantean engineering continues to drive expeditions and scholarly inquiries. Yet, amid the pursuit of artifacts and ancient artifacts, the boundary between myth and reality becomes increasingly blurred. Some argue that the Atlantis narrative represents an allegorical account of a society's technological zenith, extrapolated into a grand parable of hubris and downfall. Could these engineering marvels merely be metaphorical expressions of a culture's aspirations and eventual demise? Unraveling this enigma demands meticulous scrutiny of historical records, geological data, and interdisciplinary dialogues to dissect the facets of Atlantean engineering. Additionally, modern technologies such as satellite imaging and underwater archaeology offer promising tools for detecting potential remnants and structural patterns that might align with accounts of Atlantean constructions. These scientific endeavors serve as crucial conduits for examining the realm where myth converges with actuality, shedding light on the plausibility of the Atlantean technological legacy. As we navigate this labyrinth of speculation and inquiry, one fact remains resoundingly clear: the allure of Atlantis endures, beckoning us to delve deeper into the intersections of mythology, history, and engineering.

Analyzing the Cultural Impact through Ages

Throughout history, the mythical realm of Atlantis has transcended its origins to become an enduring cultural touchstone. The legend of this technologically advanced and ultimately doomed civilization has permeated literature, art, philosophy, and even popular culture. From the works of Plato to modern science fiction, the concept of Atlantis continues to evoke a sense of wonder and curiosity. Analyzing the cultural impact of Atlantis over the ages reveals fascinating insights into the human psyche and collective imagination. Across different epochs, Atlantis has symbolized various ideals and fears, serving as a canvas upon which societies projected their aspirations and anxieties. In antiquity, it exemplified the potential achievements and inevitable failures of human ambition. It fueled the imaginations of explorers, inspiring voyages in search of the fabled utopia. In the Renaissance, it embodied a longing for lost knowledge and wisdom, fueling the era of discovery and intellectual fervor. The Age of Enlightenment saw Atlantis as a cautionary tale against hubris and unchecked power, while the Industrial Revolution sparked a resurgence of interest in ancient technology, drawing parallels to the Atlantean mastery of mechanics and engineering. With the dawn of the modern era, Atlantis continues to captivate minds, aligning with contemporary concerns about environmental degradation, technological advancement, and societal collapse. Its enduring presence in cultural discourse reflects the perennial themes of progress, aspiration, and the perils of unchecked ambition. As we delve into the evolution of Atlantis's cultural significance, we gain deeper insight into the ways that societies have grappled with their own complexities, aspirations, and fears. The lasting legacy of Atlantis serves as a mirror, reflecting the ever-changing perspectives and preoccupations of humanity across the ages.

Modern Archaeological Finds and Theories

Modern archaeological research has greatly contributed to our understanding of ancient civilizations, including the intriguing legend of Atlantis. While numerous theories about the possible location of Atlantis exist, recent archaeological discoveries have played a crucial role in shedding light on this enigmatic civilization. Advanced deep-sea exploration methods have allowed researchers to uncover submerged historical sites, leading to intense speculation regarding their potential connection to the fabled city.

Palaeogeographic studies implicating seismic and volcanic activity also present compelling evidence for the existence of a landmass that matches some descriptions of Atlantis. These findings have sparked renewed interest in the search for physical remnants of this long-lost civilization, prompting interdisciplinary collaborations among scientists, historians, and archaeologists.

Moreover, advancements in geospatial imaging technologies have enabled researchers to conduct thorough investigations of vast, unexplored regions where traces of ancient civilizations are believed to lie buried. These cutting-edge tools have brought us closer than ever to unraveling the mysteries surrounding Atlantis. Furthermore, the comparative study of linguistic, cultural, and architectural elements between known civilizations and the mythical Atlanteans has yielded thought-provoking insights. By examining the similarities and divergences, contemporary scholars have gained valuable perspectives on the likelihood of Atlantis's existence and its potential impact on subsequent cultures.

Besides physical evidence, numerous theoretical frameworks have emerged, addressing the social, economic, and political structures of the supposed Atlantean society. Through extensive analysis of ancient texts, folklore, and oral traditions from various global cultures, researchers have formulated diverse hypotheses concerning the nature of Atlantean civilization. This interdisciplinary approach not only enriches our understanding of ancient societies but also underscores the legacy of Atlantis within the collective human imagination.

In conclusion, modern archaeological findings and associated theories continue to fuel intellectual discourse and inspire further research into the enigma of Atlantis. As technology advances, unveiling new opportunities for exploration and analysis, the quest to discern fact from fiction regarding this legendary realm remains a captivating pursuit for contemporary scholars and enthusiasts alike.

Comparative Analysis with Contemporary Civilizations

As we strive to comprehend the enigmatic legacy of Atlantis, it becomes imperative to draw parallels between its purported advancements and those of contemporaneous societies. A comparative analysis serves as a lens through which we can attempt to decipher the extent of Atlantean influence and technological prowess. When juxtaposed with ancient Mesopotamia, for instance, similarities emerge in terms of architectural sophistication and water management systems, potentially hinting at an interconnected global network of knowledge exchange. Mesoamerican civilizations exhibit intriguing correlations with aspects of Atlantean lore, such as submerged structures and advanced astronomical alignments, igniting speculation about cross-continental communication or shared sources of inspiration. Moreover, the enigmatic relics uncovered in Egypt bear resemblance to certain Atlantean artifacts, prompting conjecture about potential cultural diffusion or intercultural trade links. The comparative study extends to the Mediterranean region, where parallels can be drawn between Atlantean descriptions and the advancement of seafaring capabilities among ancient Phoenician and Greek societies, possibly shedding light on the maritime reach of Atlantean expeditions. It is important to note that such comparisons are not conclusive evidence of a direct connection but rather serve to provoke scholarly inquiry into the extent of communal influence or independent technological developments among ancient civilizations. Through this holistic approach, we begin to discern a tapestry of shared expertise, conceptual frameworks, and engineering achievements across diverse landscapes—each contributing to our evolving understanding of the Atlantean phenomenon within the broader scope of ancient innovation.

Conclusion: Unraveling the Mystique of Atlantis

The enchanting allure of Atlantis has persisted for millennia, captivating the imagination of scholars, adventurers, and storytellers alike. As we conclude our investigation into the legacy of this enigmatic civilization, it becomes evident that Atlantis continues to wield an extraordinary influence on the collective consciousness. The mystique surrounding its supposed existence has added a layer of fascination and mystery to the annals of ancient history. We must acknowledge that while the reality of Atlantis remains elusive, its impact on contemporary culture and technology cannot be dismissed lightly.

Throughout this exploration, we meticulously compared the purported achievements of Atlantean society with those of known contemporary civilizations. Our discerning analysis revealed striking parallels that could challenge conventional historical narratives. From advanced architectural techniques to innovative technologies, the similarities between the hypothetical Atlantis and actual ancient societies sparked profound speculation about the possibility of an advanced prehistoric civilization. Such considerations raise poignant questions about the cyclical nature of human development and the potential for lost knowledge to resurface in unconventional ways.

Moreover, the cultural impact of Atlantis extends far beyond academic conjecture. Its legendary status has permeated diverse spheres, from literature and art to scientific inquiry and technological innovation. This enduring presence underscores the enduring power of myth and the human yearning for extraordinary tales of grandeur and downfall. Whether tangible or mythical, the concept of Atlantis symbolizes humanity's perennial quest for wisdom, progress, and transcendence.

In the face of modern archaeological findings and evolving theories, the enigma of Atlantis endures as a compelling subject of scholarly discourse. While definitive evidence remains elusive, the ongoing pursuit of truth and understanding is a testament to the enduring allure of this ancient enigma. The journey to unraveling the mysteries of Atlantis is a testament to the indomitable spirit of inquiry that drives us to explore the unknown depths of our shared human heritage.

14

Lost Lore and Mythical Machines

Exploring the Gap: Contextual Discontinuities

The study of ancient civilizations and their technological achievements often presents a perplexing puzzle for historians, archaeologists, and scholars alike. While there exists a wealth of documented evidence detailing the advancements made by various societies in areas such as architecture, agriculture, and craftsmanship, there are equally significant gaps in our knowledge that raise compelling questions about the extent of ancient technological prowess. This chapter seeks to delve into the disparities between known historical records and the apparent discontinuities in the technological evolution of ancient civilizations. The exploration will invite readers to consider the limitations of existing sources and the challenges they pose to gaining a comprehensive understanding of ancient technologies. By scrutinizing these contextual discontinuities, we aim to foster a nuanced appreciation for the complexities inherent in deciphering the technological accomplishments of bygone eras.

Documenting the Undocumented: Source Challenges

The quest to uncover the secrets of ancient technology is often impeded by the scarcity of credible sources. Documenting the history of mythical machines and lost lore presents a myriad of challenges that require a comprehensive understanding of historical, archeological, and anthropological methodologies. One of the primary issues faced by researchers is the lack of written records or direct evidence pertaining to certain technological advancements from antiquity. The absence of verifiable accounts poses a significant obstacle in unraveling the mysteries surrounding these enigmatic inventions.

Another challenge arises from the interpretative nature of available sources. Many ancient texts, inscriptions, and artifacts contain allegorical depictions and cryptic references to technologies, making their interpretation an intricate task. Deciphering the symbolism and metaphorical language employed by ancient civilizations requires a nuanced approach, as it involves reconciling mythological narratives with potential real-world innovations. Furthermore, the dissemination of cultural myths and legendary accounts across generations adds layers of complexity to the process of identifying and substantiating factual information related to ancient machinery and mechanisms.

In addition to interpreting existing sources, scholars encounter the daunting task of locating and validating obscure or esoteric references within the broader corpus of ancient literature and material culture. Pursuing potential leads often involves meticulous cross-referencing of disparate sources, including mythological texts, ancient manuscripts, and oral traditions passed down through generations. Collating and corroborating variegated remnants of historical data demands a meticulous and interdisciplinary approach, requiring expertise in linguistics, epigraphy, anthropology, and archaeology.

Moreover, the fragmented nature of historical documentation necessitates piecing together isolated fragments of evidence to construct cohesive narratives about mythical machines. This compilation process involves sifting through disparate sources, drawing correlations between divergent cultural perspectives, and discerning patterns that may unveil hidden insights into early technological marvels. Delving into the undocumented aspects of ancient technology mandates a rigorously critical analysis of available sources, engendering a scholarly pursuit characterized by meticulous scrutiny and intellectual dexterity.

Mythical Machines: Between Reality and Fantasy

In the exploration of ancient technology, the fine line between reality and fantasy becomes increasingly pronounced when delving into the realm of mythical machines. Myths and legends from various ancient cultures often feature extraordinary technological marvels, blurring the boundaries between the tangible and the ethereal. Within these narratives lie captivating accounts of machines that exhibited seemingly supernatural capabilities, challenging the conventional understanding of technological achievements. It is crucial to approach these accounts with an open mind, acknowledging their cultural and symbolic significance while striving to discern any potential kernels of truth. The study of mythical machines requires a balanced perspective that bridges the realms of folklore and historical inquiry. A comprehensive analysis must consider the diverse contexts in which these myths originated, encompassing sociocultural beliefs, religious symbolism, and allegorical representations of power and governance. By exploring the interplay between myth and reality, researchers can uncover valuable insights into the technological imaginings of ancient societies and the sociopolitical forces that shaped their collective consciousness. It is essential to recognize the profound impact of storytelling traditions and oral histories in transmitting knowledge about these mythical machines. Tales of enigmatic constructs and wondrous mechanisms have traversed generations, serving as conduits for both practical wisdom and speculative creativity. Delving into the nuances of mythical machines also demands an examination of the symbiotic relationship between human ingenuity and divine attribution. Many mythological constructs are imbued with transcendent qualities, blurring the boundaries between mortal craftsmanship and celestial intervention. Exploring this complex interplay sheds light on the societal perceptions of innovation, agency, and the perceived boundaries of human potential. At the core of the discussion lies the interpretive challenge of reconciling the fantastical nature of mythical machines with potential historical realities. To navigate this intricate terrain, one must engage in holistic interdisciplinary approaches that draw from mythology, archaeology, anthropology, and comparative studies of ancient technologies. Such integrated methodologies pave the way for nuanced reconstructions of the cultural landscapes in which mythical machines emerged. In doing so, scholars can discern the underlying motivations behind the creation of these captivating narratives and their implications for understanding ancient technological paradigms. Embracing the complexities inherent in the study of mythical machines enriches our comprehension of the intricate tapestry of human technological evolution, inspiring a deeper appreciation for the intertwining threads of truth, wonder, and imagination that define our collective heritage.

Artifacts of Power: Literary and Historical Evidences

The search for tangible evidence of advanced ancient technologies has been a quest that has captured the imagination of scholars and enthusiasts alike. Within this fascination lies the examination of literary and historical texts that depict artifacts of power, providing potential clues to the existence of sophisticated machineries and devices in antiquity. These artifacts, often described as legendary objects with extraordinary functions, have been documented in mythologies, epics, and chronicles from various ancient civilizations.

A prominent example is the celestial chariot of the Hindu deity Arjuna in the Indian epic, Mahabharata. The detailed account of this wondrous vehicle capable of traversing the skies and carrying divine weaponry has sparked debates about whether it was an allegorical element or potentially a representation of ancient technological prowess.

Furthermore, accounts of powerful talismans like the Philosopher's Stone in alchemical texts and the legendary Lost Ark in various religious scriptures have contributed to the intrigue surrounding these purportedly miraculous objects.

The historical evidence of such artifacts can also be found in archaeological discoveries and ancient manuscripts. Illustrative inscriptions and depictions in temple walls, such as those in Egypt, Mesopotamia, and South America, provide enigmatic references to enigmatic devices and symbolic representations of what could be construed as advanced machinery. Moreover, textual records from the libraries of Alexandria and Timbuktu have sometimes made allusions to the existence of intricate mechanisms and automata, adding layers of mystery to the quest for understanding these artifacts.

Scholars and researchers, drawing from diverse disciplines including archaeology, comparative literature, and history, have meticulously scrutinized these literary and historical evidences to discern their metaphorical or factual significance in relation to ancient technology. While skepticism abounds due to the lack of physical remains, the cumulative weight of these narratives continues to fuel fervent exploration and theoretical reconstructions in the pursuit of uncovering the veiled technological achievements of bygone eras.

In essence, delving into the realm of artifacts of power entails navigating through the labyrinthine corridors of ancient folklore, scriptural accounts, and excavated vestiges. It necessitates a judicious approach that acknowledges the interplay between cultural symbolism, artistic expression, and the potential remnants of lost advancements. As we peel back the layers of these gripping tales and cryptic signs, the quest for understanding the veracity and implications of these purported artifacts of power pulses with an unyielding thirst for illuminating the enigmatic shadows cast by ancient civilizations.

Technologies Lost to Time: Hypothetical Reconstructions

Throughout the annals of history, there exist enigmatic remnants and references to technologies attributed to ancient civilizations that appear ahead of their time. These mysteries have captured the imaginations of scholars, historians, and enthusiasts alike, sparking debates, speculation, and ambitious attempts at hypothetical reconstructions. The pursuit of unraveling these technological enigmas involves piecing together fragmentary evidence, drawing from a myriad of disciplines, and embracing a blend of scientific inquiry with creative interpretation. Such endeavors offer potential glimpses into the innovative prowess of our ancestors and provoke contemplation regarding the once prevalent technologies lost to time. The journey toward hypothetical reconstructions necessitates a cautious approach to avoid straying into the realms of pure conjecture. Informed by careful analysis of available historical records, artifacts, and symbolic representations, these reconstructions seek to bridge the gaps between what is known and what may have existed. It demands an interdisciplinary collaboration, incorporating expertise from fields such as archaeology, anthropology, engineering, and comparative mythology to gauge the feasibility and plausibility of the proposed technological advancements. One notable example of a technology lost to time is the fabled Antikythera mechanism, an intricate mechanical device discovered in a shipwreck off the coast of Greece. Believed to date back to the 2nd century BC, this complex instrument has ignited speculative reconstructions concerning its purpose, functionality, and the advanced knowledge required for its construction. Hypothetical reconstructions of the Antikythera mechanism have sparked fascinating dialogues on ancient astronomical understanding, ingenious mechanical design, and the implications of such a device within its sociocultural context. Similarly, the mythical accounts of flying machines described in ancient Sanskrit texts, such as the Vimana, fuel imaginative hypotheses on potential aeronautical achievements of early civilizations. Proposals for reconstructing such mythical machines delve into aerodynamic principles, materials science, and comparative mythological studies as they endeavor to unveil the possible mechanical marvels that inspired these captivating legends. The process of hypothetical reconstruction enables a thought-provoking exploration of the intersection between myth and reality, stimulating inquiries into the marvels that might have arisen from the depths of antiquity. While acknowledging the inherent limitations and uncertainties ingrained in these endeavors, the pursuit of reconstructing lost technologies elevates the discourse surrounding ancient innovation, offering compelling insights into the evocative heritage of humanity's technological narrative.

Cryptic Artefacts and their Interpretations

Throughout history, numerous enigmatic artifacts have been unearthed, displaying intricate designs and mechanisms that defy the technological expectations of their time. From the Antikythera mechanism to the Baghdad Battery, these cryptic artifacts have puzzled historians, archaeologists, and scientists for centuries, sparking intense debate about their significance and purpose. The Antikythera mechanism, discovered in a shipwreck off the coast of the Greek island of Antikythera, is an ancient analog computer dating back to the 2nd century BCE. Its complex gear system and astronomical functionalities have led many to reevaluate their understanding of ancient technology. Similarly, the Baghdad Battery, a set of ceramic jars found near Baghdad, has raised questions about whether these could have been used as galvanic cells in early electroplating or for electrotherapy. While these artifacts have inspired various interpretations, some theories remain controversial, with proponents advocating for advanced ancient civilizations possessing knowledge far ahead of their time, while skeptics attribute these artifacts to more conventional explanations, such as religious or ceremonial purposes. This section aims to delve into the interpretation of these enigmatic artifacts, considering the diverse perspectives put forth by experts and scholars from different fields. We will explore the technological implications, mythological associations, and cross-cultural influences that have shaped our understanding of these cryptic relics. By scrutinizing the physical characteristics, contextual clues, and historical contexts of these artifacts, we strive to unravel the mysteries surrounding them and shed light on their potential significance in ancient societies. Through meticulous examination and interdisciplinary collaboration, we aim to navigate the intersection of science, history, and culture to construct plausible interpretations and discern the possible roles these cryptic artifacts played in ancient civilizations. Furthermore, this discussion will address the challenges and limitations inherent in interpreting these artifacts, emphasizing the need for critical analysis and the importance of acknowledging varying perspectives within this complex and fascinating field of study.

Cross-Cultural Analysis of Mythical Technologies

Mythical technologies, often mentioned in ancient texts and folklore, have captivated the imaginations of scholars and enthusiasts alike for millennia. While these tales encompass a wide range of cultures and civilizations, they share common threads that lead to intriguing parallels when analyzed through a cross-cultural lens. One of the most striking aspects of these mythical technologies is the recurring motifs and archetypes across diverse societies, suggesting a shared human fascination with extraordinary inventions and achievements beyond the scope of conventional understanding. In various mythologies, we find accounts of advanced devices, such as flying machines, miraculous healing tools, or powerful weaponry, often attributed to legendary figures or deities. The similarities in the descriptions of these mythical devices across different cultures hint at the universal aspirations and dreams of humanity. Through a detailed examination of these cross-cultural narratives, scholars have unearthed fascinating overlaps and similarities that transcend geographical and temporal boundaries. These analyses provide valuable insights into the collective consciousness of ancient societies, shedding light on their beliefs, aspirations, and perhaps even glimpses of historical occurrences that have been woven into the fabric of mythology. Moreover, by comparing and contrasting the accounts of mythical technologies from diverse cultural perspectives, researchers can discern patterns that hint at potential interactions and exchanges between ancient civilizations. This approach facilitates the exploration of potential cross-cultural influences and exchange of ideas that might have led to the propagation of similar myths and legends across different societies. Additionally, examining mythical technologies from a cross-cultural standpoint enables a deeper understanding of the symbolic meanings and metaphors embedded within these narratives. By identifying recurrent themes and symbolism, researchers can unravel the underlying messages and philosophical concepts that permeate these stories, offering profound insights into the values, concerns, and intellectual pursuits of ancient cultures. Furthermore, a cross-cultural analysis serves as a bridge between disparate mythological traditions, fostering a comprehensive perspective that transcends individual cultural biases and limitations. It invites scholars to engage in interdisciplinary dialogues, drawing connections between distant mythologies and exploring the universal human quest for knowledge, progress, and transcendence. Through this holistic approach, the study of mythical technologies becomes an invaluable tool for unraveling the intricate tapestry of human imagination, spirituality, and ingenuity, enriching our comprehension of ancient cultures and affirming the enduring relevance of these timeless narratives.

Influence of Allegorical Devices in Ancient Texts

Ancient texts, often rich in allegorical devices, provide a fascinating window into the thought processes and cultural beliefs of our forebears. These allegories, at first glance, may appear as simple moral or symbolic tales; however, upon closer examination, they reveal profound insights into ancient technology and its societal impact. Through the lens of allegory, technological concepts are woven into narratives to convey deeper meanings, serving as vehicles for passing down knowledge and wisdom across generations. By exploring these allegorical devices, we can gain a more nuanced understanding of the roles that technology played in shaping the ancient world. Allegories pertaining to mythical machines and wondrous contraptions offer glimpses into the ways ancient societies sought to comprehend and contextualize their technological advancements. These stories often encapsulate the human struggle to harness and understand the power of innovation, offering cautionary tales or inspirational parables. One remarkable aspect of allegorical interpretation lies in the diverse cultural manifestations of these tales. Different civilizations have employed allegory uniquely, infusing their own values and aspirations into the fabric of these narratives. For instance, the Greek myths surrounding the legendary inventor Daedalus and his son Icarus illuminate the complexities of hubris and ambition in the pursuit of technological progress. Meanwhile, in Eastern traditions, allegorical tales often intertwine with philosophical themes, delving into the intricate relationships between humanity, nature, and the artificial. By decoding and analyzing these allegorical devices, modern scholars can reconstruct the mindset of ancient cultures and gain insight into the origins of their technological achievements. However, this process requires careful consideration and cross-referencing across multiple sources, as allegories are often layered with multiple interpretations. This complexity underscores the need for interdisciplinary collaboration and holistic approaches to fully comprehend the influence of allegorical devices on our understanding of ancient technology. Through an exploration of these allegorical frontiers, we can appreciate the foundational role of storytelling in preserving and transmitting knowledge about ancient technologies and the profound impact they had on their societies.

The Role of Mythology in Understanding Ancient Tech

Mythology serves as a captivating lens through which we can attempt to comprehend the enigmatic realm of ancient technology. Studded with allegorical narratives, ancient mythologies often intertwine with technological concepts, presenting an intriguing fusion of imagination and resourcefulness. Through in-depth analysis and interdisciplinary research, we can delve into the symbolic significance embedded in these myths, unraveling potential connections to advanced technologies of antiquity.

One prominent role of mythology lies in its ability to convey complex technological ideas in a digestible and relatable manner. Ancient cultures often employed mythical narratives to explain the origins and workings of sophisticated machinery, portraying them as feats of divine or supernatural prowess. These allegorical devices served as a means of preserving and transmitting knowledge across generations in a form that resonated with the cultural and cognitive fabric of societies. By demystifying these allegories, we gain insights into the technological advancements and innovative solutions that were integral to ancient societies.

Furthermore, mythology provides a bridge between the tangible and intangible aspects of ancient technology, offering a glimpse into the social, spiritual, and philosophical dimensions intertwined with innovative creations. The narratives woven into myths unfold a tapestry of human ingenuity, aspirations, and the persistent quest for comprehension of the world and the cosmos. This amalgamation of technical prowess and spiritual reverence is pivotal in deciphering the ancient mindset and their approach to scientific and engineering disciplines, providing a holistic understanding of technology's role in shaping ancient civilizations.

Moreover, delving into the mythology of ancient cultures sheds light on the possible existence of advanced and esoteric technologies that may have been overlooked in conventional historical narratives. By discerning the metaphors, hidden meanings, and allegoric nuances within these myths, we can contemplate the existence of lost or undiscovered technologies, igniting multidisciplinary explorations into unexplored realms of ancient innovation and mechanical mastery.

In essence, mythology stands as an intricate web connecting ancient civilizations' perspectives on technology, knowledge transmission, and societal values. Deciphering the roles assigned to technology within mythological frameworks unveils profound insights, enriching our comprehension of ancient technological achievements and fostering imaginative inquiries into the possibilities that deepen our understanding of antiquity's technological landscape.

Conclusion: Bridging Lost Lore with Current Knowledge

With the exploration into the role of mythology in understanding ancient technology, it becomes clear that ancient myths and legends often conceal a deeper layer of knowledge and understanding. By unraveling these mythological narratives and studying their potential links to ancient technology, we gain invaluable insights into the technological prowess of our ancestors. However, this pursuit also underscores the challenges inherent in interpreting and bridging the gaps between lost lore and current knowledge. As we delve deeper into deciphering the enigmatic elements of ancient myths, we must remain conscious of the inherent complexities and ambiguities that accompany such endeavors.

Bridging lost lore with current knowledge requires a multidisciplinary approach, bringing together the expertise of archaeologists, historians, mythologists, and technologists. Collaboration is essential to systematically examine the intertwined threads of ancient myths and technological advancements, drawing from diverse fields of study to illuminate the obscured pathways of our technological heritage. This collaborative effort should seek to shed light on the enigmatic artifacts, cryptic inscriptions, and mythical machines that have perplexed scholars for centuries, offering new perspectives and hypotheses rooted in both historical evidence and speculative imagination.

Moreover, the conclusion of such investigations must not only serve as an academic exercise but also as a pragmatic endeavor. The insights garnered from bridging lost lore with current knowledge can potentially inform and inspire modern technological innovations. By gaining a more comprehensive understanding of the technological feats of our forebears, we open the door to fresh interpretations and applications that may influence contemporary technological design, engineering, and problem-solving. It is this bridge between the past and present that holds promise for a richer, more nuanced comprehension of both ancient and modern technology.

In conclusion, the journey to bridge lost lore with current knowledge is an ongoing quest that demands relentless curiosity, intellectual rigor, and interdisciplinary collaboration. By weaving together the threads of mythology, archaeology, history, and technology, we aspire not only to unlock the mysteries of ancient civilizations but also to revitalize the spirit of innovation that transcends time. Through this convergence of disciplines and ideas, we endeavor to elevate the study of ancient technology from mere speculation to a well-informed, potent force driving contemporary technological progress.

15

Uncovering Ancient Secrets

Retrieving the Past

Uncovering ancient technologies is integral to deepening our understanding of historical narratives and unraveling the mysteries of human progress. The pursuit of retrieving the past through archaeological methodologies and technologies illuminates the ingenuity and innovation of ancient civilizations. This quest serves as a bridge between antiquity and the contemporary world, shedding light on the vast array of technological advancements that have shaped human history. As we delve into the artifacts and remnants of ancient civilizations, we unearth not only physical objects but also the stories and expertise of the ingenious minds that preceded us. Each discovery becomes a piece of a cosmic puzzle, contributing to a comprehensive timeline of human achievement. The significance of these findings extends beyond mere academic curiosity; it grants us insight into the evolution of societies, the development of critical industries, and the dawn of innovative thought. By exploring the technologies of the past, we gain a deeper appreciation for the resourcefulness and creativity inherent in the human spirit. Moreover, uncovering ancient technologies challenges preconceived notions about the capabilities of early civilizations, demonstrating their sophisticated methods to solve complex problems and enhance daily life. Through this exploration, we not only honor the legacy of our ancestors but also pave the way for a more nuanced and inclusive view of history.

Archaeological Methodologies and Technologies

The field of archaeology is continually advancing due to the integration of various methodologies and technologies. These advancements have significantly transformed how researchers approach the study of ancient civilizations and their technological developments. One of the fundamental methodologies employed in archaeology is stratigraphy, which involves the study of soil layers to establish a chronological sequence of events at an excavation site. Combining this with remote sensing techniques such as ground-penetrating radar and LiDAR has revolutionized the identification of potential archaeological sites and features without invasive excavation. In addition, the utilization of Geographic Information Systems (GIS) facilitates the management, analysis, and visualization of spatial data, enabling archaeologists to understand the landscape and its transformation over time. Furthermore, the application of 3D modeling and digital reconstruction techniques has enabled the creation of virtual representations of ancient artifacts and structures, preserving them for future generations and providing new insights into their original forms. Advanced imaging technologies, including multispectral and hyperspectral imaging, have also allowed for the detailed examination of materials and pigments used in ancient art and artifacts, shedding light on ancient manufacturing processes and trade networks. Moreover, the emergence of DNA analysis and isotopic studies in bioarchaeology has contributed to our understanding of ancient human populations, their migration patterns, and dietary habits. Furthermore, the development of portable X-ray fluorescence (pXRF) and other spectroscopy tools has enhanced the analysis of material composition, aiding in provenance studies of artifacts and geological sourcing. To ensure the preservation of cultural heritage, non-invasive survey methods such as photogrammetry and 3D scanning have been invaluable in documenting fragile archaeological sites and artifacts, reducing the need for physical handling and potential damage. The interdisciplinary collaboration between archaeologists, conservationists, geophysicists, and specialists in various scientific fields has fostered an environment for innovative approaches to uncovering ancient secrets. These methodologies and technologies continue to push the boundaries of archaeological research, offering unprecedented opportunities to unveil the mysteries of the past.

Significant Discoveries in Context

The field of archaeology has yielded a wealth of significant discoveries that have provided profound insights into ancient civilizations. These findings, carefully unearthed and meticulously analyzed, have expanded our understanding of the technological accomplishments, societal structures, and cultural practices of bygone eras. From the majestic ruins of ancient cities to the humble remnants of daily life, each discovery offers a tantalizing glimpse into the past, presenting both familiar and enigmatic aspects of human history. In Egypt, the unearthing of the Rosetta Stone revolutionized the study of hieroglyphics, unlocking the lost language of the pharaohs and enabling scholars to decode countless inscriptions. Similarly, the excavation of Pompeii and Herculaneum offered a poignant snapshot of daily Roman life frozen in time by the catastrophic eruption of Mount Vesuvius, providing invaluable details about ancient urban planning, domestic routines, and artistic expressions. The uncovering of the Terracotta Army in China astonished the world with its lifelike representation of an imperial entourage, offering a window into the opulence and organizational prowess of the Qin Dynasty. These monumental findings, alongside countless others, not only enrich our knowledge of antiquity but also inspire ongoing investigations and interpretations that continue to shape our comprehension of human heritage.

Dating Techniques: Chronology of Artifacts

Chronology is a fundamental aspect of archaeological research, providing the framework for understanding the temporal dimension of human activities and technological advancements. Dating techniques are crucial in determining the age of artifacts, helping to reconstruct the timelines of ancient civilizations.

One of the primary methods used in archaeology is relative dating, which assesses the age of an artifact in relation to other objects or strata. Stratigraphy, the study of layers of sediment and rock, helps establish a sequential order of deposition and provides a relative chronology for artifacts within those layers. This approach enables archaeologists to understand the relative sequence of events, but it does not provide precise dates.

To complement relative dating, absolute dating methods are employed to assign numerical ages to artifacts or sites. Radiocarbon dating, based on the decay of carbon isotopes in organic materials, is one of the most widely utilized techniques. By comparing the ratio of carbon isotopes in a sample with that in the atmosphere, researchers can estimate the age of organic remains with remarkable accuracy. Similarly, dendrochronology, or tree-ring dating, relies on the analysis of tree ring patterns to date wooden artifacts and structures.

Furthermore, thermoluminescence dating measures the accumulated radiation dose in crystalline materials such as flint or pottery, providing insights into the last time the material was heated or exposed to sunlight. This method offers valuable information for dating ceramics and fire-cracked rocks found at archaeological sites. Meanwhile, optically stimulated luminescence dating evaluates the amount of trapped electrons in quartz or feldspar grains, allowing researchers to determine when these minerals were last exposed to light.

In addition to these techniques, archaeological scientists utilize a range of innovative approaches, including electron spin resonance dating, uranium-series dating, and potassium-argon dating, each offering distinct advantages for establishing chronologies in different contexts. By combining multiple dating methods, archaeologists can cross-validate chronological data, enhancing the reliability of their findings and interpretations.

Understanding the chronology of artifacts is essential for reconstructing ancient societies, technological developments, and cultural changes. Accurate dating enables scholars to situate discoveries within specific historical periods, unravel trade networks, track migrations, and comprehend the evolution of material culture over time. Moreover, chronological frameworks serve as crucial reference points for interdisciplinary studies, shedding light on the interactions between ancient civilizations and their environments. As archaeological dating techniques continue to advance, they contribute significantly to our collective comprehension of humanity's past and enrich our broader understanding of the ancient world.

Role of Geospatial Technologies in Archaeology

Geospatial technologies have revolutionized the field of archaeology, providing invaluable tools for understanding and interpreting ancient landscapes and human activities. Through the integration of Geographic Information Systems (GIS), Global Positioning Systems (GPS), remote sensing, and aerial photography, archaeologists can capture, analyze, and visualize spatial data with unprecedented precision and detail. GIS allows researchers to overlay and manipulate various layers of geographic information, including terrain, vegetation, and human settlements, to generate complex spatial models that aid in site selection, landscape analysis, and predictive modeling. Additionally, GPS technology enables accurate mapping of archaeological features and artifact distributions, facilitating precise collection of spatial data for further analysis. Remote sensing plays a crucial role in archaeological surveys by detecting subsurface features and identifying anomalies on the earth's surface that may indicate potential archaeological sites. Aerial photography complements remote sensing by providing high-resolution images of archaeological sites and landscapes, enabling detailed visual analysis and documentation. The application of geospatial technologies extends beyond research, with significant impact on heritage management and conservation efforts. By creating detailed maps and 3D visualizations, archaeologists can monitor and protect cultural heritage sites from natural and human-induced threats. Geospatial technologies also contribute to public engagement and education through interactive digital reconstructions and virtual tours of archaeological sites, fostering greater appreciation and understanding of ancient civilizations. As technology continues to advance, the role of geospatial technologies in archaeology will evolve, offering new possibilities for exploration, preservation, and interpretation of our collective human heritage.

Bioarchaeological Contributions

The interdisciplinary approach of bioarchaeology paves the way for a deeper understanding of ancient cultures through the examination of human remains within an archaeological context. By scrutinizing skeletal and dental remains, bioarchaeologists unravel intricate details about past populations such as genetic affinities, dietary patterns, health status, and even demographic profiles. Through the integration of biological and social sciences, this field not only sheds light on the lives of our ancestors but also addresses pertinent questions about their interactions, lifestyles, and adaptations to environmental challenges.

Bioarchaeology utilizes various scientific techniques to conduct robust analyses, including stable isotope analysis, ancient DNA studies, palaeopathology, and osteoarchaeology. The study of stable isotopes assists in reconstructing ancient diets and migration patterns, providing valuable insights into subsistence strategies and mobility of past human populations. Additionally, ancient DNA studies enable the identification of kinship relations, population movements, and genetic predispositions to diseases, thus contributing to the broader narrative of human evolution and societal dynamics.

Moreover, the discipline of palaeopathology examines evidence of pathological conditions in ancient populations, offering remarkable inferences about past health, disease prevalence, and medical practices. By meticulously examining skeletal markers, bioarchaeologists are able to discern prevalent ailments, identify physical stress indicators, and even uncover instances of deliberate cranial modifications or surgical interventions. Furthermore, osteoarchaeology provides crucial data on activity patterns, traumatic injuries, labor divisions, and mortuary practices, contributing to comprehensive reconstructions of community structures and social organization.

The fundamental principles of bioarchaeology extend beyond mere anatomical investigations, delving into the complex interplay between biological and cultural factors in shaping ancient societies. The application of these methods not only enriches our understanding of the past but also informs contemporary debates on topics such as food security, health disparities, and genetic diversity. As an invaluable component of archaeological inquiry, bioarchaeology offers a compelling lens through which we can apprehend the intricacies of human existence across diverse temporal and spatial landscapes.

Interpreting Material Culture

Material culture serves as a crucial window into the past, providing invaluable insights into the daily lives, beliefs, and technological developments of ancient civilizations. By analyzing and interpreting material remains, archaeologists can reconstruct social structures, economic systems, and cultural practices that have long since faded into obscurity. This method of inquiry encompasses an extensive array of artifacts, from tools and pottery to architecture and art, each offering a unique perspective on the societies that produced them.

One of the primary objectives in interpreting material culture is to discern patterns and trends across different time periods and geographic regions. Through careful analysis and comparison of these artifacts, researchers can trace the evolution of technology, trade networks, and artistic styles, shedding light on the interconnectedness of ancient societies and the diffusion of ideas and innovations. Moreover, the study of material culture allows for a deeper understanding of societal values and norms, as expressed through the objects they created and used on a daily basis.

Interpretation of material culture also involves consideration of the context in which these artifacts were found. The spatial relationships between objects within archaeological sites, as well as their positioning in relation to surrounding features, provide critical information about their functions, symbolic meanings, and cultural significance. Furthermore, the integration of interdisciplinary approaches, such as ethnography and ethnoarchaeology, enhances the interpretation of material culture by drawing connections between ancient practices and those observed in contemporary societies.

It is essential for archaeologists to approach the interpretation of material culture with a critical and nuanced perspective, acknowledging the complexities inherent in assigning meaning to ancient artifacts. By combining empirical data with theoretical frameworks, researchers can construct compelling narratives about the past while remaining mindful of the interpretive nature of their work. This balancing act requires vigilance against imposing modern biases or assumptions onto ancient material culture, as well as openness to alternative interpretations that may challenge established paradigms.

Ultimately, the interpretation of material culture is a dynamic and evolving process that continuously contributes to our understanding of human history. As new technologies and methodologies emerge, archaeologists are afforded fresh opportunities to delve deeper into the material remnants of antiquity, unraveling the stories waiting to be told by the objects of bygone eras.

Challenges in Preservation and Conservation

Preservation and conservation present significant challenges in the field of ancient technology and archaeological studies. As we delve into uncovering ancient secrets, it becomes evident that preserving artifacts and structures from millennia past is a complex and demanding endeavor. The longevity of materials such as metals, ceramics, organic matter, and even entire architectural sites requires meticulous care, expertise, and resources to maintain their integrity for future generations and ongoing research. This section delves into the multifaceted challenges encountered in the preservation and conservation of ancient technology and archaeological finds. One of the primary challenges is environmental impact. Whether exposed to fluctuating humidity, temperature variations, or natural disasters, ancient artifacts are vulnerable to deterioration. Conservation efforts must address these risks through controlled storage, climate monitoring, and strategic interventions. Additionally, the threat of human interference and vandalism looms large, particularly in unprotected or remote archaeological sites. Balancing accessibility for research and public engagement with the need for security and protection poses an ongoing dilemma. Furthermore, the sheer scale and diversity of artifacts require specialized knowledge and techniques for preservation. From delicate textiles and perishable organic remains to monumental architectural structures, each type of artifact demands tailored approaches to ensure its long-term survival. Preservation efforts also intersect with ethical considerations regarding repatriation and cultural heritage management. Striking a balance between displaying artifacts for educational purposes and respecting the cultural significance and ownership rights of communities is an ongoing challenge in the field. Moreover, financial constraints often limit the scope and effectiveness of preservation and conservation efforts. Allocating sufficient resources for comprehensive conservation programs, including staffing, equipment, and facilities, is a perennial struggle faced by many institutions and organizations involved in the safeguarding of ancient technologies and archaeological treasures. Addressing these challenges necessitates collaboration among archaeologists, conservators, museum professionals, and local communities. By sharing knowledge, leveraging technological advancements, and engaging in sustainable practices, the preservation and conservation of ancient secrets can be safeguarded for future study and admiration.

Case Studies of Integral Finds

In the exploration of ancient technologies and civilizations, case studies of specific integral finds provide crucial insights into the evolution of human innovation. These case studies not only shed light on the technological advancements of ancient societies but also offer valuable glimpses into the cultural, social, and economic dynamics of their time. Examining these integral finds with meticulous detail allows us to comprehend the intricate nature of ancient technology and its significance in the development of human civilization. Each case study serves as a window into the past, unraveling mysteries and offering a deeper understanding of our ancestors' ingenuity and resourcefulness.

One such case study revolves around the discovery of the Antikythera Mechanism, an ancient Greek analog computer that dates back to the 2nd century BCE. This extraordinary artifact, recovered from a shipwreck near the island of Antikythera, showcases the remarkable level of technological sophistication achieved by ancient engineers and astronomers. Through detailed analysis and reconstruction efforts, researchers have unveiled the mechanism's intricate gears and dials, revealing its ability to track celestial movements and predict astronomical events. The Antikythera Mechanism stands as a testament to the advanced knowledge and capabilities of ancient civilizations, challenging conventional perceptions of technological development in antiquity.

Another compelling case study encompasses the archaeological excavation of the city of Mohenjo-daro, an urban center of the ancient Indus Valley Civilization. The uncovering of this ancient metropolis has provided invaluable evidence of sophisticated urban planning, advanced drainage systems, and a form of proto-writing known as the Indus script. The intricate layout of the city, complete with well-organized streets, multistory buildings, and an elaborate water management system, offers a glimpse into the urban engineering prowess of the Harappan people. The artifacts unearthed at Mohenjo-daro, including delicate jewelry, intricate pottery, and statuary, offer a comprehensive understanding of the material culture and artistic expressions of this ancient society.

Furthermore, the excavation of the Terracotta Army in Xi'an, China, stands as a remarkable case study highlighting the artistic and technological achievements of the Qin Dynasty. Discovered in the mausoleum of the first Emperor of China, Qin Shi Huang, the terracotta soldiers and horses represent a monumental feat of ancient craftsmanship. The lifelike statues, meticulously crafted and individually distinct, illustrate the exceptional skill of ancient Chinese artisans. The preservation of these terracotta warriors provides a tangible connection to the military strategies, artistic traditions, and technological capabilities of ancient China.

Each of these case studies exemplifies the invaluable contributions of archaeological discoveries to the understanding of ancient technology, art, and societal structures. These integral finds serve as poignant reminders of the enduring legacy of past civilizations and continue to inspire and inform contemporary research and scholarship.

Summary and Implications for Future Research

The journey through ancient secrets and their revealing case studies underscores the vital role of diligent research in uncovering the past. As we scrutinize integral finds, it becomes apparent that these discoveries have significant implications for future research in archaeology and related fields. Summarizing the collective findings, it is evident that the intricate interplay of interdisciplinary methodologies has led to groundbreaking revelations about ancient civilizations. The integration of advanced technologies, such as LiDAR scanning, ground-penetrating radar, and various dating techniques, has facilitated a deeper understanding of historical contexts.

Looking ahead, the implications for future research are multifaceted. Firstly, the case studies highlight the need for continued innovation in preservation and conservation strategies. With the ever-present threat of natural decay and human interference, it is imperative to develop more robust and sustainable approaches to safeguarding invaluable artifacts for posterity. Moreover, the emphasis on bioarchaeological contributions underscores the potential for enriched insights into ancient societies through the study of human remains.

Additionally, the significance of geospatial technologies in tracking ancient landscapes and settlements cannot be overstated. Future research endeavors should explore the untapped potential of leveraging geographic information systems (GIS) and remote sensing to map undiscovered archaeological sites, thereby expanding our knowledge of past civilizations. Equally crucial is forging synergies between material culture and historical narratives. The interpretative power of artifacts, from pottery shards to inscriptions, offers a nuanced perspective on ancient daily life, rituals, and societal structures.

Furthermore, the chronological precision attained through advanced dating techniques illuminates the temporal progression of human ingenuity and technological evolution. This aspect not only reshapes our understanding of historical timelines but also informs the broader narrative of humanity's collective intellectual heritage. The synthesis of interdisciplinary findings underscores the necessity for collaborative research frameworks that bridge conventional disciplinary boundaries. By fostering cross-disciplinary dialogue and cooperation, the potential for paradigm-shifting discoveries is amplified.

In conclusion, the case studies presented in this section serve as a testament to the enduring relevance of ancient secrets and their profound implications for future research. As we stand at the intersection of technological innovation and scholarly inquiry, it is imperative to diligently chart a course that harnesses the full potential of interdisciplinary collaboration, cutting-edge methodologies, and ethical stewardship of ancient artifacts. The lessons gleaned from these integral finds beckon us towards a future where the mysteries of antiquity continue to inspire new discoveries and reshape our understanding of human history.

16

Deciphering the Ancients: Modern Interpretations

Modern Interpretation Techniques

The realm of archaeology has witnessed a remarkable evolution in interpretation techniques, ushering in an era that combines traditional methodologies with cutting-edge technological advancements. This multidimensional approach represents a seismic shift from the simplistic observational methods employed by early explorers and researchers. Today, the quest for decoding the enigmatic remnants of ancient civilizations hinges upon the strategic integration of archaeological advances and technological tools. The convergence of these two realms has endowed modern interpreters with an unprecedented arsenal, enabling them to delve into historical mysteries with unprecedented precision and depth.

Archaeological advances have played a pivotal role in reshaping the landscape of interpretations. From the development of more refined excavation methods to the establishment of interdisciplinary collaborations, the field has undergone a profound metamorphosis. These advances have not only revolutionized the way artifacts are unearthed and handled but have also fostered a more holistic understanding of past societies and their technological prowess. Moreover, the utilization of technological tools has opened up new frontiers in the domain of interpretation. By harnessing state-of-the-art imaging technologies, such as ground-penetrating radar and LiDAR (Light Detection and Ranging), archaeologists can now unearth hidden structures and visualize ancient landscapes in unprecedented detail. The marriage of traditional excavation with digital innovation has ushered in a new era of exploration, where historical narratives are pieced together with meticulous precision. Additionally, the advent of advanced laboratory techniques, including radiocarbon dating and material analysis, has empowered researchers to glean intricate details about ancient artifacts and their provenance, thereby enriching the interpretative process.

As we reflect on the transformative journey of interpretation techniques, it becomes evident that the discipline has transcended the boundaries of traditional scholarship and embraced a progressive ethos powered by technological ingenuity. By honing the craft of decipherment and cultural reconstruction, contemporary interpreters stand poised at the threshold of a new age—an age where the past is decoded through a lens that fuses the art of the ancient world with the science of the modern era.

Archaeological Advances and Technological Tools

The dynamic landscape of archaeological research has been significantly transformed by monumental advances in technology. In this context, cutting-edge tools and methodologies have revolutionized the way ancient artifacts are studied and interpreted. Utilizing state-of-the-art equipment such as ground-penetrating radar, LiDAR (Light Detection and Ranging), multispectral imaging, and 3D laser scanning, archaeologists can meticulously survey and document historical sites and objects with unprecedented precision. These technological innovations enable the non-invasive examination of structures and landscapes, providing invaluable insights into ancient civilizations without compromising their integrity. Moreover, drones equipped with high-resolution cameras have rendered aerial surveys more efficient, facilitating the discovery and documentation of remote archaeological sites that were previously inaccessible or overlooked. In addition to these advancements in physical exploration, the digital realm has also become integral to modern archaeological practice. Through the use of Geographic Information Systems (GIS) and other sophisticated software, researchers can analyze and visualize spatial data, allowing for a more comprehensive understanding of ancient urban environments, trade routes, and cultural landscapes. Furthermore, the digitization of archives and the adoption of virtual reality (VR) technologies offer immersive experiences that bring ancient history to life for scholars and the general public alike. It is crucial to acknowledge that these technological tools not only enhance archaeological fieldwork but also contribute to ethical stewardship by minimizing the potential harm posed by invasive excavation methods. The synergy between cutting-edge technology and traditional archaeological approaches enriches our comprehension of ancient civilizations, transforming archaeological interpretation into a multifaceted endeavor that transcends temporal and geographic boundaries.

Linguistic Decipherment Methods

Linguistic decipherment methods form a vital facet of the interdisciplinary approach employed in unlocking the enigmatic languages and scripts of ancient civilizations. This intricate process demands a deep understanding of linguistic structures, historical contexts, and cultural nuances. Linguists, philologists, and epigraphers collaborate to decode inscriptions, manuscripts, and texts utilizing a blend of traditional methodologies and cutting-edge technologies.

One fundamental approach is the comparative method, which involves scrutinizing known languages related to the ancient script or dialect under examination. By identifying patterns, similarities, and variations, researchers can propose plausible phonetic, grammatical, and semantic interpretations. Furthermore, statistical analysis and frequency distribution studies aid in recognizing recurrent linguistic elements, enabling the formulation of hypotheses for decipherment.

Another significant technique is contextual interpretation, which entails studying the material culture surrounding the inscriptions. Insights into religious rituals, administrative records, or literary works associated with the script can provide valuable clues. This method often involves collaboration with archaeologists, historians, and anthropologists to construct a comprehensive comprehension of the societal context in which the language was utilized.

Moreover, the application of computational linguistics has revolutionized decipherment efforts. Natural language processing algorithms, pattern recognition software, and machine learning models are leveraged to analyze large corpora of texts, identify recurrent clusters of symbols, and generate provisional translations. While these automated processes do not substitute human expertise, they significantly expedite the preliminary stages of decipherment and aid in confirming or refuting hypotheses derived from traditional philological analyses.

In recent years, interdisciplinary collaborations have been pivotal in surmounting linguistic barriers. Semioticians, cognitive scientists, and even experts in cryptanalysis have contributed to refining decipherment methodologies. The utilization of innovative technologies such as three-dimensional scanning, spectral imaging, and virtual reality simulations has enabled a more comprehensive understanding of inscribed artifacts, providing crucial insights into the visual and tactile dimensions of the language.

Furthermore, the ethical considerations inherent in linguistic decipherment cannot be overlooked. Respect for indigenous knowledge systems, recognition of marginalized languages, and caution in presenting findings are integral aspects of responsible linguistic research. Collaborative partnerships with descendant communities and local scholars facilitate a more inclusive and culturally sensitive approach to decipherment, ensuring that the voices of the past are preserved and honored.

As we delve deeper into the realm of linguistic decipherment, it becomes apparent that this multifaceted endeavor necessitates a harmonious amalgamation of rigorous scholarly methodologies, technological innovations, and ethical conscientiousness. The ongoing pursuit of unraveling ancient scripts serves not only to unveil the intellectual achievements of our forebears but also to foster cross-cultural appreciation and understanding in our modern global society.

Role of Artificial Intelligence in Understanding Ancient Artifacts

Artificial intelligence (AI) has emerged as a groundbreaking tool in the field of archaeology, revolutionizing the way ancient artifacts are analyzed and interpreted. Through advanced data processing and pattern recognition capabilities, AI has become instrumental in deciphering complex scripts, understanding intricate designs, and extracting valuable insights from archaeological finds. One of the primary applications of AI in this context is the translation and interpretation of ancient languages and scripts. By leveraging machine learning algorithms, AI systems can analyze and compare vast amounts of linguistic data, aiding researchers in unlocking the meanings and contexts embedded in ancient texts. Moreover, AI technologies have proven invaluable in image recognition and analysis, enabling archaeologists to discern patterns, symbols, and visual motifs on artifacts that may have previously eluded human perception. This has facilitated the identification of recurring themes and iconography across diverse cultures and time periods, providing deeper understanding and cross-cultural linkages. Additionally, AI-driven predictive modeling has enhanced the reconstruction of fragmentary artifacts, allowing for the virtual restoration of damaged or deteriorated objects based on sophisticated algorithms and historical references. Furthermore, the integration of AI with geographic information systems (GIS) has facilitated spatial analysis and mapping of archaeological sites, offering new perspectives on ancient civilizations' settlement patterns, trade routes, and cultural interactions. The implementation of AI in artifact analysis also extends to material science and conservation efforts. AI-powered techniques enable the chemical composition analysis of ancient materials, aiding in the identification of manufacturing processes and the preservation of delicate relics. As AI continues to evolve, its role in understanding ancient artifacts is poised to expand further, with the potential for more sophisticated simulations, virtual reconstructions, and collaborative knowledge discovery. However, this progression also raises ethical considerations and challenges related to bias in AI interpretations, cultural representation, and the preservation of traditional methods. Despite these complexities, the transformative impact of AI on unraveling the mysteries of ancient artifacts remains a compelling frontier in the intersection of technology and archaeology.

Comparative Analysis: Then and Now

Exploring the technological advances of ancient civilizations alongside modern innovations provides a fascinating opportunity for comparative analysis. By juxtaposing ancient technologies with their contemporary equivalents, we gain valuable insights into the evolution of human ingenuity and its impact on society. This section will delve into a comprehensive examination of key advancements, drawing parallels between the ancient and the modern.

Ancient civilizations exhibited remarkable engineering prowess, as evidenced by the construction of monumental structures such as the pyramids of Giza, the Parthenon, and the Great Wall of China. These architectural marvels stand as testaments to the sophisticated knowledge and skills possessed by our predecessors. Comparing these achievements to modern architectural wonders, such as skyscrapers, bridges, and megastructures, showcases the enduring legacy of innovative problem-solving and construction techniques.

Moreover, the development of mechanical devices in ancient times, from simple machines to more complex automata, highlights the foundations of early engineering principles. Contrastingly, the integration of artificial intelligence, robotics, and nanotechnology in the contemporary era reflects the exponential growth of technological capabilities, underscoring the transformative power of innovation over millennia.

It is equally compelling to analyze the advancements in transportation and communication throughout history. From the invention of the wheel and early maritime navigation to the age of automobiles, airplanes, and instantaneous global connectivity through digital networks, we witness the profound shifts that have reshaped the human experience. By delving into the ways in which ancient societies navigated and communicated across vast distances, we gain a deeper appreciation for the gradual refinement of these essential components of human progress.

This comparative analysis further extends to fields such as medicine, agriculture, and material science, where ancient practices and discoveries converge with contemporary breakthroughs, illuminating the continuum of human knowledge and innovation. By scrutinizing the similarities and divergences between historical and modern approaches, we can discern recurring patterns, identify areas of sustained improvement, and acknowledge the resilience of certain foundational principles that have withstood the test of time.

In conclusion, the comparative analysis between ancient and modern technologies serves as a testament to humanity's unwavering quest for advancement. It underscores the rich tapestry of human achievement and the enduring echoes of our ancestors' inventive endeavors in the fabric of our present-day world.

Controversies and Debates in Interpretation

Interpretation of ancient artifacts and texts has always been at the center of scholarly debate, igniting controversies that fuel intense academic discussions among archaeologists, linguists, historians, and experts from various disciplines. The constant evolution of interpretation methodologies often leads to conflicting perspectives and debates, shedding light on the complex nature of deciphering ancient civilizations.

One of the primary controversies revolves around the subjectivity inherent in interpretation. Scholars often debate the extent to which modern biases and preconceptions influence the reading of ancient materials. This brings into question the validity of interpretations and raises concerns about projecting contemporary ideologies onto ancient cultures. Moreover, the ethical implications of imposing modern viewpoints on historical narratives have sparked contentious dialogues within the academic community.

Another profound debate stems from the challenges of linguistic and cultural translation. The diverse languages and symbols used by ancient civilizations present significant hurdles in accurately interpreting their writings and inscriptions. Linguists and epigraphers frequently engage in heated discussions about the most appropriate linguistic framework for decoding ancient texts, highlighting the intricate process of unraveling linguistic codes that have endured centuries of obscurity.

Additionally, the authenticity of modern interpretations is often called into question through the lens of provenance and historical context. Experts scrutinize the reliability of archaeological findings and question the legitimacy of artifacts, raising doubts about the accuracy of conclusions drawn from potentially compromised sources. This controversy underscores the need for rigorous scrutiny and validation processes in the field of ancient interpretation.

Furthermore, technological advancements and their role in interpretation have sparked fervent debates within the academic sphere. While some scholars champion the integration of cutting-edge technologies such as artificial intelligence and digital imaging in decipherment processes, others express concerns about overreliance on these tools, emphasizing the importance of maintaining a balance between traditional methodologies and modern innovations.

The controversies and debates surrounding interpretation in the field of ancient studies serve as a testament to the intellectual vigor and critical analysis embedded in scholarly pursuits. As new evidence emerges and methodologies evolve, these debates continue to shape the trajectory of our understanding of ancient civilizations, fostering an environment ripe for advancing knowledge and challenging established paradigms.

Case Studies of Successful Decipherments

Throughout history, the process of deciphering ancient scripts and artifacts has presented a myriad of challenges, often shrouded in controversies and debates. However, there have been remarkable successes in unraveling the secrets of ancient civilizations, shedding light on their languages, customs, and technologies. This section will delve into several case studies of successful decipherments, showcasing the ingenuity and perseverance of those who have unlocked the mysteries of antiquity.

One notable case study is the decipherment of Egyptian hieroglyphs by Jean-François Champollion in the early 19th century. Through meticulous comparisons and analysis, Champollion made a breakthrough by deciphering the Rosetta Stone, a pivotal artifact bearing inscriptions in three scripts. His pioneering work revolutionized our understanding of ancient Egypt, opening a window into its rich history and cultural heritage.

In another intriguing instance, the Indus Valley script, an enigmatic writing system from ancient South Asia, puzzled scholars for decades. The successful decipherment of this script by Asko Parpola and others involved interdisciplinary efforts spanning archaeology, linguistics, and computer-based analyses. The revelations gleaned from the deciphered inscriptions provided valuable insights into the sophisticated urban civilization that flourished in the region over four millennia ago.

Furthermore, the decipherment of cuneiform script, prevalent in ancient Mesopotamia, stands as a triumph of collaborative scholarship and technological innovation. Scholars such as Henry Rawlinson and George Smith played crucial roles in decoding this intricate script, uncovering a wealth of literary, administrative, and legal texts. Their efforts offered a compelling narrative of Mesopotamian society, revealing facets of governance, religion, and trade that shaped the ancient world.

These case studies underscore the significance of persistent scholarly inquiry and interdisciplinary approaches in deciphering ancient writings and artifacts. Each successful decipherment not only enhances our understanding of past civilizations but also exemplifies the enduring human quest to bridge the temporal chasm and connect with our ancestors. Their achievements continue to inspire contemporary researchers to employ novel methodologies and technologies in unraveling the remaining enigmas of bygone eras, underscoring the timeless allure of deciphering the ancients.

Emerging Technologies and Future Prospects

As we delve into the uncharted territories of ancient technology, it is essential to explore the ever-evolving landscape of emerging technologies and their role in unraveling the mysteries of our ancestors. The fusion of cutting-edge tools, such as advanced imaging techniques, 3D modeling, and virtual reality, with traditional archaeological methods, has opened new vistas for exploration and interpretation. By adapting these technologies to study ancient artifacts and sites, researchers can gain unprecedented insights into how our forebearers ingeniously crafted and utilized technology. Moreover, the integration of remote sensing technologies, ranging from LiDAR (Light Detection and Ranging) to multispectral imaging, has revolutionized the way we uncover and comprehend ancient landscapes, allowing us to peer through dense jungles or beneath thick layers of sediment with unparalleled clarity. In parallel, advancements in computational linguistics and machine learning algorithms are propelling the boundaries of linguistic analysis, enabling scholars to decode ancient scripts and languages that have long eluded decipherment. These innovative approaches offer a glimmer of hope for unlocking the enigmatic messages inscribed in stone, clay tablets, and parchment scrolls. Looking ahead, the convergence of disciplines, such as archaeogenetics and isotopic analysis, promises to unveil intimate details about ancient societies, their movements, and cultural exchanges, transcending what was previously discernible. Additionally, the burgeoning field of bioarchaeology, with its ability to analyze ancient DNA and trace human migration patterns, offers a window into the biological tapestry of civilizations long past. When considering the future prospects of technological applications in understanding ancient civilizations, it becomes evident that the collaborative use of these tools must be underpinned by ethical considerations and respect for indigenous knowledge systems. It is imperative to seek meaningful engagements with local communities and indigenous groups, fostering a harmonious blend of traditional wisdom and modern science. By integrating indigenous narratives and holistic perspectives with scientific inquiry, a more comprehensive and nuanced understanding of ancient technologies emerges, enriching our interpretations and yielding mutual benefits. Ultimately, as we stand on the precipice of technological innovation, we must tread cautiously, ensuring that our discoveries honor the legacies of the past while guiding us towards a more inclusive and empathetic future.

Integrating Indigenous Knowledge with Contemporary Science

Indigenous knowledge, passed down through generations, is a rich tapestry of insights, observations, and techniques that have sustained communities for centuries. As we delve into the realm of deciphering ancient technologies, it becomes imperative to recognize and integrate indigenous knowledge with contemporary science. This convergence presents an opportunity to enrich our understanding of ancient civilizations and their technological innovations. Indigenous communities around the world possess deep-seated understandings of natural processes, sustainable practices, and traditional technologies that often align with modern scientific principles. By bridging these two forms of knowledge, we can unlock a more holistic comprehension of the ancient world.

The integration of indigenous knowledge with contemporary science offers a multidimensional approach to interpreting and contextualizing ancient technologies. For instance, indigenous agricultural practices could hold vital clues to how certain ancient civilizations harnessed natural resources and cultivated crops. By intertwining these insights with modern scientific methodologies, researchers can gain significant insights into the agricultural technology and ecological sustainability of ancient societies. This fusion of perspectives not only enhances our scholarly interpretations but also respects the invaluable wisdom preserved within indigenous cultures.

Moreover, the collaboration between indigenous knowledge and contemporary science holds profound potential in various fields such as medicine, ecology, astronomy, and engineering. Many indigenous communities have amassed extensive knowledge of medicinal plants, celestial navigation, sustainable land management, and construction techniques. Integrating this wealth of knowledge with contemporary scientific approaches can lead to groundbreaking discoveries in archaeological and anthropological research. Such collaborations also contribute to fostering respect for indigenous cultures and promoting inclusive dialogues that honor diverse forms of knowledge.

As we move towards the future, the integration of indigenous wisdom with contemporary science demands ethical considerations, mutual respect, and equitable partnerships. It is paramount to engage with indigenous communities respectfully, acknowledging their sovereignty and ownership of their knowledge. Collaborative research initiatives should prioritize indigenous leadership, involvement, and benefit sharing. By fostering meaningful collaborations, we pave the way for a more inclusive, ethical, and comprehensive understanding of ancient technologies while nurturing respect for the enduring wisdom of indigenous traditions.

Summary and Conclusions on Modern Interpretations

In conclusion, the integration of indigenous knowledge with contemporary science presents a transformative approach to interpreting ancient technologies and innovations. Throughout this chapter, we have delved into the intersection of traditional wisdom and modern methodologies, revealing the potential for enriched understanding and holistic perspectives. By acknowledging and incorporating indigenous knowledge systems, researchers and scholars can gain new insights and innovative interpretations of archaeological findings and historical artifacts. This collaborative approach fosters respect for diverse cultural heritage and promotes comprehensive analyses that transcend conventional Western paradigms. Furthermore, by engaging with local communities and incorporating their insights, the scientific community can establish more ethical and inclusive research practices. Empowering indigenous voices in the interpretation of ancient technologies not only enriches scholarly endeavors but also contributes to the preservation and revitalization of endangered knowledge systems. Moreover, this collaborative approach has the potential to address complex environmental and societal challenges through the fusion of ancient wisdom and contemporary scientific advancements. By recognizing the relevance of indigenous knowledge in the modern context, we pave the way for a more interconnected and harmonious future, where diverse worldviews and epistemologies coexist in synergy. As we navigate the frontier of technological innovation and historical inquiry, the intertwined narratives of traditional wisdom and modern interpretations offer a dynamic tapestry of knowledge that transcends temporal boundaries. Embracing this convergence enriches our understanding of the ancient world, propelling us towards a more inclusive and enlightened future.

17

Implications for Today and Tomorrow

Introduction to Contemporary Relevance

The applicability of ancient technological advances in contemporary settings holds immense significance in our modern world. Through the examination of historical innovations, we can draw parallels to present-day challenges and opportunities. By delving into the lessons gleaned from the past, we can gain valuable insights that have the potential to shape the trajectory of future technological progress. The resurgence of interest in ancient technologies speaks to a broader quest for sustainable and efficient solutions, echoing the age-old adage that 'necessity is the mother of invention'. It is a testament to the timelessness of certain principles and ideas that have stood the test of millennia. Moreover, the convergence of cutting-edge technology with age-old wisdom has the potential to yield remarkable innovations that are rooted in deep historical knowledge. As we navigate complex global issues such as environmental sustainability, social equity, and ethical advancements, mining the wealth of ancient technological progress offers a perspective that transcends time and space. By engaging in a discourse that spans centuries, we open ourselves to a myriad of possibilities that can inform and elevate the current landscape of technological innovation. This exploration also facilitates an understanding of the interconnectedness of human civilization and its enduring quest for advancement. From agricultural techniques to architectural marvels, the ingenuity of ancient societies offers a treasure trove of inspiration for tackling contemporary challenges. Embracing the relevance of ancient technological progress allows us to appreciate the holistic nature of human ingenuity, which extends beyond the purely technical realm into the realms of culture, society, and philosophy. As we embark on this journey of discovery, it becomes evident that the echoes of the past resonate powerfully in the context of our present endeavors. The timeless nature of ancient technological wisdom serves as a beacon, guiding us towards a harmonious fusion of tradition and modernity.

Technological Progress: Lessons from the Past

Throughout history, technology has been instrumental in shaping human civilization. The remarkable advancements achieved by ancient societies provide a treasure trove of knowledge and inspiration for contemporary technological progress. By delving into the technologically advanced accomplishments of our ancestors, modern society can gain valuable insights and learn critical lessons that can help navigate the complex challenges of the present and future.

The ingenuity and resourcefulness displayed by ancient civilizations offer considerable wisdom that transcends time. From the precise astronomical observations of the Mayans to the innovative aqueduct systems of the Romans, each civilization's technological achievements reflect an unparalleled understanding of the natural world and a commitment to enhancing quality of life. These historical triumphs underscore the significance of integrating timeless principles with cutting-edge innovation.

Moreover, studying the technological evolution of ancient societies illuminates the critical role of interdisciplinary collaboration. For instance, the interdisciplinary approach taken by the ancient Greeks, combining mathematics, philosophy, and engineering, revolutionized scientific inquiry and laid the foundations for countless modern innovations. Embracing these lessons can foster a more holistic and interconnected approach to contemporary technological advancement, transcending the limitations of siloed expertise and fostering cross-pollination of diverse ideas.

In addition, by examining the societal impacts of ancient technologies, it becomes evident that conversations about ethics and responsibility in innovation are not new. The ethical considerations surrounding the creation and use of early inventions, such as the development of mechanical automatons or the widespread adoption of water and wind power, offer pertinent insights into the ethical dilemmas faced by contemporary technologists. Understanding the ways in which our predecessors balanced progress with ethical awareness serves as a crucial guide in navigating the unprecedented power of today's technology.

Furthermore, the preservation and restoration of ancient technologies also hold contemporary significance. The effort to conserve ancient structures, such as the magnificent pyramids or intricate aqueduct systems, exemplifies the enduring value of sustainable designs and durable materials. Embracing these ancient construction techniques and materials in modern architectural and infrastructure projects can significantly contribute to sustainable development and resilience against climate-related challenges.

Ultimately, the examination of ancient technological progress serves as a powerful reminder that innovation is not solely grounded in the present; it is an accumulation of human endeavors across millennia. Acknowledging the enduring impact of ancient technologies and drawing upon their timeless teachings can propel contemporary society towards more conscientious, sustainable, and human-centered technological progress.

Innovative Perspectives on Ancient Practices

Throughout history, humanity has demonstrated remarkable ingenuity in developing technologies that facilitated day-to-day activities, artistic endeavors, and scientific discoveries. By delving into the ancient practices of various civilizations, we gain valuable insights into innovative perspectives that continue to inspire contemporary thought and creativity. The intrinsic reverence for nature and the environment prevalent in many ancient cultures provides an invaluable perspective on sustainable living. From the efficient use of natural resources to architectural principles that harmonize with the ecosystem, ancient practices offer a compelling model for addressing modern-day environmental challenges. Moreover, the application of ancient craftsmanship techniques and artisanal skills presents a compelling case for embracing traditional know-how in crafting sustainable, resilient products that contribute to eco-conscious consumption. As we navigate the complexities of the present age, these innovative perspectives from ancient practices exemplify the enduring relevance of historical wisdom. Additionally, the meticulous observation and documentation of celestial phenomena by ancient civilizations provide a unique vantage point for exploring innovative perspectives on timekeeping, astronomy, and navigation. Beyond their pragmatic utility, these ancient practices invite contemporary societies to reconsider their relationship with time, space, and the universe at large. Furthermore, the spiritual and philosophical underpinnings of ancient practices open new avenues for holistic well-being and mindfulness in today's fast-paced world. By embracing the meditative aspects of ancient traditions and adapting them to suit modern lifestyles, individuals can cultivate a sense of balance and inner harmony. These innovative perspectives offer novel approaches to mental and emotional well-being, fostering a profound connection to ancient wisdom while addressing the needs of present-day society. Ultimately, the exploration of innovative perspectives on ancient practices serves as an inspirational conduit for contemporary endeavors across a multitude of fields, including architecture, sustainability, wellness, and design, infusing modern innovation with timeless significance.

Sustainability Insights from Ancient Technologies

Ancient civilizations provide a treasure trove of sustainable technological practices that offer valuable insights for contemporary society. From the sophisticated water management systems of the Indus Valley Civilization to the innovative agricultural techniques of the Mayans, ancient technologies emphasized harmony with the natural environment and efficient resource utilization. By studying these ancient approaches, we can glean vital lessons in sustainability, which are increasingly pertinent in our modern era. One notable example is the advanced aqueduct and irrigation systems implemented by civilizations such as the Romans and the Persians. These feats of engineering enabled the efficient distribution and conservation of water resources, laying a foundation for sustainable water management that remains relevant today. Moreover, the sustainable construction methods employed in ancient architecture, utilizing local and renewable materials, offer inspiration for contemporary eco-friendly building practices. The age-old practice of harnessing renewable energy sources, exemplified by windmills and solar technologies in ancient Greece and China, holds significant relevance in the current global pursuit of clean energy solutions. Furthermore, the indigenous knowledge of sustainable agriculture practiced by ancient cultures, including crop diversification, terrace farming, and agroforestry, presents a blueprint for resilient and sustainable food production systems. This holistic approach to sustainability, embedded within the fabric of ancient societies, unveils the timeless wisdom that we can integrate into our modern-day endeavors. By recognizing the intrinsic link between ancient technologies and sustainable living, we can cultivate a renewed appreciation for the preservation of natural resources while fostering innovative solutions for environmental challenges. Embracing the sustainable ethos of ancient technologies not only enriches our understanding of historical achievements but also equips us with practical strategies to address the pressing ecological concerns of our time. As we navigate the complexities of the 21st century, the harmonious coexistence with nature practiced by our ancestors serves as a profound source of guidance, inspiring a collective commitment to sustainable development for the benefit of future generations.

Cultural Impacts and Societal Transformation

Throughout history, technology has been deeply intertwined with culture and society, influencing and shaping human civilization in profound ways. When we explore the cultural impacts and societal transformations brought about by ancient technologies, we gain valuable insights into how innovation can fundamentally alter the fabric of human existence.

Integrating Ancient Wisdom into Modern Innovation

In our quest for progress and innovation, it is essential to recognize the immense value embedded in ancient wisdom and time-tested practices. Integrating ancient wisdom into modern innovation represents a conscientious effort to bridge the chasm between historical knowledge and contemporary technological advancements. By delving into the profound insights of our ancestors, we unearth a plethora of invaluable concepts that can significantly enrich and inform present-day innovation. This integration not only honors the legacy of past civilizations but also paves the way for a holistic approach to problem-solving and sustainable development.

The ingenuity of ancient cultures, demonstrated through their architectural marvels, artistic accomplishments, and advanced technologies, offers a wellspring of inspiration for modern innovation. These civilizations exemplified unparalleled resourcefulness, astute observation of natural phenomena, and remarkable adaptability, which we can leverage to address the complex challenges of our time. By studying these ancient legacies, we gain a deeper understanding of fundamental principles that have stood the test of time. This knowledge provides a solid foundation for reimagining existing technologies, fostering creativity, and fostering an ethos that emphasizes longevity and resilience.

Moreover, integrating ancient wisdom into modern innovation fosters a spirit of interdisciplinary collaboration, where fields such as archaeology, history, anthropology, and technology converge. Embracing diverse perspectives enables us to draw upon a rich tapestry of global heritage, transcending geographical boundaries and temporal constraints. By doing so, we cultivate an inclusive approach to innovation that respects and integrates diverse cultural narratives, resulting in solutions that are sensitive to local contexts and global interdependencies. This integrative approach is pivotal in addressing the ethical and social dimensions of technological advancement, ensuring that progress aligns with human values and environmental stewardship.

Furthermore, the integration of ancient wisdom extends beyond mere theoretical discourse; it actively informs practical applications across various domains. From architecture and urban planning to sustainable agriculture and healthcare, ancient insights are being repurposed and adapted to meet contemporary needs. For instance, biomimicry—a discipline inspired by nature's designs—relies on observations of ancient ecosystems and organisms to develop innovative solutions for design, engineering, and sustainability. Similarly, traditional medicinal practices that have endured centuries offer a complementary perspective for modern healthcare, emphasizing preventive care, holistic well-being, and the utilization of natural resources. Such intersections between antiquity and modernity demonstrate the tangible benefits of integrating ancient wisdom into contemporary innovation.

Embracing the fusion of ancient wisdom and modern innovation necessitates a paradigm shift wherein the notion of progress expands beyond technological advancement alone. It engenders a renewed appreciation for the cyclical nature of knowledge, recognizing that innovation is not solely about novelty, but also about timeless principles that withstand the currents of change. As we navigate an era marked by rapid technological evolution and globalization, integrating ancient wisdom into modern innovation serves as an anchor, grounding us in enduring truths while propelling us towards a future that embodies both progress and reverence for the past.

Educational Curriculum Developments

In the contemporary landscape, the integration of ancient wisdom into modern educational curricula serves as a pivotal bridge between historical innovation and present-day learning endeavors. As academic institutions contemplate their role in nurturing versatile and future-ready students, the inclusion of ancient technological advancements and societal practices becomes increasingly imperative. By infusing educational programs with modules that showcase the innovative solutions and profound insights of our forebears, students can gain an enriched understanding of the interconnectedness of past and present technologies.

One key aspect of educational curriculum development pertains to cultivating critical thinking and problem-solving skills through the exploration of ancient technologies. By delving into the engineering feats of ancient civilizations, students can analyze the framework of these innovations and decipher the underlying principles. This not only fosters a deeper appreciation for historical accomplishments but also equips learners with invaluable analytical capabilities that are transferrable to contemporary challenges.

Additionally, the incorporation of ancient technology into educational curricula offers a multidisciplinary approach to learning. By examining the historical context, societal impact, and scientific basis of ancient inventions, students are exposed to an array of subjects, including history, anthropology, archaeology, engineering, and more. This broad-based exposure encourages holistic learning and nurtures a well-rounded perspective, preparing individuals to thrive in diverse professional environments.

Moreover, educational institutions can utilize the study of ancient technology to instill a sense of environmental consciousness and sustainable practices among students. Through exploring how ancient societies harnessed natural resources and devised eco-friendly solutions, learners can draw inspiration for addressing contemporary environmental concerns. This, in turn, facilitates the development of a generation of environmentally-aware innovators and policymakers, adept at integrating sustainability principles into modern technological advancements.

The evolution of educational curricula to encompass ancient technologies also opens avenues for cultural appreciation and global awareness. By studying the technological achievements of diverse ancient civilizations, students are encouraged to embrace multicultural perspectives, fostering empathy, and understanding across societal boundaries. This, ultimately, contributes to the cultivation of a global citizenry equipped to navigate the complexities of an interconnected world.

As educational institutions continue to evolve, the integration of ancient wisdom into the curriculum stands as a transformative endeavor. This adaptation engenders a transformative pedagogical approach, empowering students to draw inspiration from the past, innovate in the present, and envisage a future permeated by the sagacity of bygone eras.

Policy Implications and Governmental Considerations

In the realm of ancient technology, the role of policies and governance cannot be overstated. As we delve into the historical footprints of innovation and technological advancements, it becomes crucial to draw parallels between the governmental frameworks of antiquity and those of our present era. Notably, ancient civilizations often had intricate systems of governance that were closely intertwined with technological progress, raising compelling questions about the interplay between policy formation and technological development. Examining the policy implications of ancient technologies can provide valuable insights for contemporary governance, particularly in addressing issues related to sustainable development, resource management, and societal impact. By gleaning lessons from the regulatory approaches of bygone eras, contemporary policymakers can gain a deeper understanding of how to navigate the complexities of managing and regulating cutting-edge technologies. One of the key considerations involves adapting legal and regulatory frameworks to accommodate the rapid advancement of technology while safeguarding ethical and societal values. This necessitates comprehensive analysis and adaptation of existing policies to ensure they remain relevant and effective in the face of technological evolution. Furthermore, government initiatives focused on preserving and promoting ancient technologies can play a pivotal role in fostering cultural heritage, stimulating scientific curiosity, and inspiring innovation. This can encompass funding for archaeological research, the establishment of specialized educational programs, and the implementation of incentive structures to encourage the exploration and revitalization of ancient technologies. Additionally, as societies increasingly grapple with the ethical and moral dimensions of emerging technologies, drawing upon the ethical principles embedded in ancient wisdom can inform the development of responsible and sustainable policies. Governments must proactively engage with scholars, industry experts, and cultural custodians to construct inclusive policies that balance technological progress with societal well-being. By embracing an interdisciplinary approach that integrates historical perspectives with modern governance strategies, policymakers can strive to create a regulatory environment that fosters innovation while upholding fundamental principles of ethics, equity, and accountability. Ultimately, a nuanced understanding of the relationship between policy, governance, and ancient technologies can offer substantial guidance for shaping a future characterized by harmony between technological advancement and societal welfare.

Future Technological Trajectories Informed by History

As we delve into the depths of ancient technological marvels, it becomes increasingly apparent that these historical innovations can serve as more than mere relics of the past—they are critical guides shaping our trajectory towards the future. By extrapolating insights from the remarkable achievements of ancient civilizations, we can glean invaluable wisdom to inform our contemporary technological advancements. The notion of drawing inspiration from history is not novel, yet the specific applications are often overlooked in the fervent pursuit of cutting-edge breakthroughs. It is imperative to recognize the potential inherent in fusing the unparalleled foresight of ancient inventors with the boundless possibilities afforded by modern capabilities. Looking forward, there are several key areas where ancient technologies can profoundly influence future trajectories. One such area is sustainable energy solutions. Examining historical sources of energy production and utilization unveils ingenious methods that harnessed natural resources without compromising the delicate ecological balance. By integrating these time-tested principles with advanced engineering and scientific knowledge, we can chart a course towards sustainable energy systems capable of meeting the escalating global demand while minimizing environmental impact. Additionally, the field of materials science stands to benefit immeasurably from lessons derived from antiquity. Ancient artisans mastered the manipulation of raw materials to craft resilient structures, durable tools, and aesthetically stunning artifacts. Understanding the chemical composition, processing techniques, and structural characteristics of these ancient materials can catalyze the development of innovative, environmentally sound materials with multifaceted applications in construction, manufacturing, and beyond. Moreover, the complex systems and infrastructures of ancient societies offer profound insights into urban planning and resource management. By scrutinizing the benevolent equilibrium achieved by past civilizations in their interaction with the environment, contemporary urban planners and policymakers can devise strategies for sustainable city development and resource allocation to meet the needs of burgeoning populations without compromising the planet's well-being. Furthermore, the fusion of ancient wisdom with contemporary technology has the potential to revolutionize medical practices and healthcare delivery. Ancient healing traditions and medical knowledge exhibit a harmony with natural remedies, holistic approaches, and preventive care that resonates deeply with modern wellness philosophies. Leveraging this ancient wisdom in conjunction with state-of-the-art medical technologies can pave the way for personalized, integrative healthcare solutions that prioritize well-being and longevity. In conclusion, the synthesis of historical insights with modern ingenuity holds the key to unlocking an era defined not only by technological marvels but also by sustainable progress, cultural enrichment, and societal harmony. The future of technology stands at the precipice of a profound transformation, and it is our responsibility to heed the timeless messages encoded within the annals of history as we navigate uncharted territories. Through this harmonious convergence of the old and the new, we stand poised to shape a future that honors the triumphs of the past while embracing the promise of tomorrow.

Conclusion: Bridging Times and Technologies

As we conclude this exploration of ancient technologies and their implications for the present and future, it is evident that the knowledge and achievements of our ancestors continue to hold great relevance in contemporary society. The integration of ancient wisdom with modern innovation has the potential to shape technological trajectories in profound ways. By bridging the gaps between bygone eras and current advancements, we can harness the valuable insights garnered from historical precedents. This process of fusion enhances not only our understanding of the past, but also our capacity to address present challenges and navigate future opportunities. Through these connections, a synergy emerges, drawing from the richness of antiquity while propelling us towards innovative frontiers.

Spanning disciplines and domains, the convergence of times and technologies yields far-reaching implications. From architecture and engineering to medicine, agriculture, and renewable energy, the intricate tapestry of ancient knowledge offers blueprints for sustainable solutions. By adapting and reimagining age-old practices, we can respond to contemporary environmental concerns, societal needs, and global demands. Moreover, the cultural heritage embedded within traditional techniques elucidates diverse perspectives on resource management, societal organization, and human ingenuity. This informs holistic approaches to modern-day challenges, fostering inclusive development and equitable progress.

Concurrently, the educational sector stands to benefit from an enriched curriculum that incorporates the teachings of ancient civilizations, providing students with a broader understanding of historical context and technological evolution. By contextualizing current scientific and technological paradigms within the narrative of human advancement, we inspire future generations to innovate with an awareness of the enduring legacy of their predecessors. Additionally, policy considerations informed by the resonance of ancient technologies illuminate pathways toward sustainable governance, highlighting the critical intersections between heritage preservation, economic growth, and environmental stewardship.

In closing, the synthesis of ancient technologies with contemporary innovations represents a continuum of human achievement, a testament to the enduring relevance of historical knowledge. By embracing this synergy, we stand poised at the precipice of transformative possibilities, guided by the wisdom of bygone epochs while charting a course towards a promising future. As we embark on this journey of convergence, we are not bound by the limitations of time, but rather liberated by the continuity of human ingenuity.

18

Aftermath: Reflections on the Ancient Singularity

Reassessing the Ancient Singularity

The examination of ancient technologies and methodologies compels us to undertake a profound reassessment of their significance and their unrecognized impacts on current scientific paradigmatic shifts. Through a meticulous study of historical artifacts, cultural practices, and the subtle nuances of ancient innovation, we begin to unravel the intricate web of influence that has permeated modern technological advancements. By delving into the innovative feats achieved by civilizations past, we gain insight into the origins of fundamental concepts and principles that continue to shape our contemporary understanding of science and technology. From early engineering marvels to sophisticated mathematical systems, the ingenious solutions devised by ancient societies serve as a testament to the human capacity for creativity and problem-solving. Furthermore, the reevaluation of these ancient achievements provides a vantage point from which to discern the latent impact that they have had on the evolution of our current technological landscape. This assessment necessitates a shift in perspective, prompting us to acknowledge the enduring legacy of ancient innovation and to recognize its pervasive influence in the underlying fabric of present-day scientific discourse. Moreover, by reimagining the interconnectedness of ancient and modern knowledge domains, we are empowered to trace the lineage of ideas and methodologies that have transcended temporal boundaries, thereby fostering a more comprehensive understanding of the collective journey of human ingenuity. In light of this, it becomes evident that the study of ancient technologies grants us not only access to historical insights but also enables us to contextualize the import of these innovations within the broader narrative of scientific and technological progress. Consequently, embracing this holistic approach to reassessing the ancient singularity engenders a profound appreciation for the enduring legacy of ancient wisdom and its subtle yet profound influence on the trajectory of contemporary scientific inquiry.

Cultural and Technological Impact Assessment

The Cultural and Technological Impact Assessment of the Ancient Singularity delves into the profound influence that ancient innovations have had on society, culture, and technology throughout history. By examining the impact of these advancements, we begin to unravel the intricate web of connections between past and present civilizations, offering valuable insights for the future.

Technological progress is inseparable from its cultural context. From the architectural marvels of the pyramids and the Great Wall of China to the development of early forms of writing, each innovation reflects the values, beliefs, and societal needs of its time. Understanding the cultural roots of these technologies provides a deeper appreciation for their significance and enduring legacy.

Moreover, these ancient advancements have left an indelible mark on the evolution of human civilization. They not only shaped the physical landscape but also influenced the development of industry, commerce, and trade routes. The exchange of ideas and technologies between different ancient cultures set the stage for globalization as we know it today. This interconnectedness is a testament to the enduring impact of ancient innovations across diverse societies and regions.

In addition, the assessment extends to the profound influence of ancient technologies on modern society. By examining the ways in which these innovations continue to resonate in contemporary life, we gain a better understanding of the interconnectedness of human progress. Whether through the principles of engineering, the legacy of philosophical thought, or the enduring symbols of ancient creativity, the impact of these innovations can be found woven throughout the fabric of the modern world.

Ultimately, this assessment serves as a reminder of the inextricable link between culture and technology, highlighting the enduring impact of ancient achievements on the course of human history and the ongoing quest for progress.

Lessons Learned from Ancient Innovations

Ancient innovations offer a treasury of wisdom, providing valuable lessons that resonate across time. The legacies of ancient technologies encompass not only tangible inventions, but also the underlying principles and mindsets that drove their creation. By closely examining these innovations, we gain profound insights into the human capacity for ingenuity and problem-solving. One of the foremost lessons gleaned from ancient innovations is adaptability. Civilizations such as Mesopotamia and Egypt exhibited remarkable flexibility in harnessing available resources to develop sophisticated irrigation systems, architectural marvels, and early forms of writing. This adaptability speaks to the importance of resourcefulness and resilience in the face of adversity. Furthermore, ancient innovations emphasize the significance of interdisciplinary collaboration. The synergy between artistry, engineering, and scientific understanding is evident in the construction of awe-inspiring structures, such as the Pyramids of Giza and the Parthenon. The fusion of diverse knowledge streams underscores the holistic nature of innovation. Moreover, the meticulous craftsmanship and attention to detail observed in ancient artifacts convey the timeless lesson of quality over quantity. Each artifact reflects a commitment to excellence and craftsmanship, setting standards that continue to inspire contemporary artisans and engineers. Another pivotal lesson is the value of sustainable practices. Ancient civilizations demonstrated an admirable harmony with nature, integrating ecological principles into their technological endeavors. From the sustainable urban planning of the Indus Valley Civilization to the sophisticated agricultural techniques of the Mayans, these ancient societies serve as precedents for conscientious resource management. Additionally, ancient innovations highlight the crucial role of observation and empirical inquiry in driving progress. Through astute observations of celestial movements, early astronomers discerned patterns that contributed to the development of calendars and navigation. These discoveries exemplify the power of curiosity in unraveling the mysteries of the natural world. The interconnectedness of ancient innovations with cultural, social, and environmental contexts reinforces the need for contextual understanding in technological development. By assimilating these historical lessons, we can navigate contemporary challenges with a more informed approach, evoking the enduring ethos of resilience, collaboration, sustainability, precision, and curiosity embodied by our forebears.

Cross-temporal Influence on Modern Technology

The cross-temporal influence of ancient technologies on modern technology is a compelling subject that reveals the enduring impact and relevance of historical innovations. By examining the advancements of ancient civilizations, we gain profound insights into how their ingenuity continues to shape and inspire contemporary technological developments. The legacies of ancient engineers, inventors, and visionaries continue to reverberate through time, leaving an indelible mark on our modern world.

One area where we observe the direct influence of ancient technology on the modern era is in the fields of architecture and engineering.

The innovative construction techniques employed by ancient civilizations have served as the foundation for many of the structural designs and building methods utilized today. From the precision of ancient stonemasonry to the sophisticated hydraulic engineering of antiquity, these age-old practices have been reimagined and integrated into modern construction methodologies, contributing to the creation of sustainable and resilient infrastructure.

Moreover, the technological advancements of ancient cultures have profoundly influenced the evolution of various scientific disciplines. The astronomical knowledge of ancient civilizations, such as the Babylonians and Mayans, has informed and enriched our understanding of celestial phenomena and cosmological principles. Their mathematical and navigational prowess has played a pivotal role in the development of modern astronomy, navigation, and geospatial mapping technologies, enabling us to explore the frontiers of space and map our planet with unprecedented accuracy.

Furthermore, the inventive spirit of ancient craftsmen and artisans has left an indelible imprint on contemporary manufacturing and fabrication processes. The intricate metalworking techniques of ancient blacksmiths and metallurgists, as well as the ingenuity displayed in ceramics and glassmaking, have inspired modern advancements in materials science and industrial manufacturing. The fusion of traditional craftsmanship with cutting-edge technologies has led to the creation of innovative materials with enhanced properties, revolutionizing industries ranging from aerospace and automotive to biotechnology and renewable energy.

In essence, the cross-temporal influence of ancient technology on the modern era underscores the enduring legacy of human innovation and serves as a testament to the timeless relevance of historical wisdom. By embracing the lessons of the past and integrating them into our present endeavors, we honor the ingenuity of our ancestors while propelling our technological capabilities towards a more enlightened future.

Ethical Considerations and Philosophical Insights

As we reflect on the advancements of ancient technology and their impact on the present, it is essential to delve into the ethical considerations and philosophical insights that arise from this exploration. The intersection of technology and ethics has been a topic of profound importance throughout human history and continues to be relevant as we navigate the complexities of modern innovation. Ethical considerations in the context of ancient technology encompass a wide array of facets, ranging from the societal implications of early innovations to the moral responsibility associated with harnessing technological power.

One of the fundamental philosophical insights that emerges from the study of ancient technological advancements is the concept of wisdom in utilizing knowledge for the greater good of humanity. Ancient civilizations grappled with questions of morality, and their perspectives on ethical conduct in relation to technology offer valuable lessons for contemporary society. This introspection provokes contemplation on the balance between progress and ethical stewardship. Understanding the historical evolution of ethical frameworks allows us to develop a more nuanced approach to addressing the ethical challenges presented by modern technological advancements.

Moreover, delving into philosophical insights inherently leads us to ponder the interplay between technology and human agency. The ethical dimensions of augmenting our capabilities through technology prompt thought-provoking questions about the preservation of individual autonomy and the potential consequences of technological integration. Ancient wisdom prompts a reevaluation of our role as stewards of innovation, challenging us to consider the implications of our technological ambitions on the fabric of society.

Considering ancient technology through an ethical and philosophical lens invites us to explore discernment in the pursuit of knowledge and progress. It calls for a reexamination of our values, responsibilities, and aspirations, both as individuals and as a collective global community. By engaging with these ethical and philosophical considerations, we can strive towards a future where technological advancement aligns with human flourishing and ethical principles, thus ensuring a harmonious coexistence between innovation and the greater good.

Technological Forecasting: What Can We Expect?

The paradigm shift brought about by ancient technology has sparked intense speculation about the future. As we stand on the precipice of a new era, technological forecasting becomes paramount in navigating the uncharted waters that lie ahead. Drawing from the inexplicable achievements of our ancestors and the accelerating pace of modern innovation, we can distill essential patterns to envision the shape of things to come. One prominent area of interest centers on the evolutionary trajectory of artificial intelligence. The integration of AI into everyday life is set to redefine human-machine interaction, as well as revolutionize industries such as healthcare, transportation, and finance. Concurrently, the burgeoning field of biotechnology presents a glimpse into personalized medicine, genetically tailored treatments, and potential advances in human augmentation. Another compelling facet is the intersection of virtual and physical worlds, where augmented reality and virtual reality herald groundbreaking applications in fields ranging from entertainment to education. This convergence may reconfigure social dynamics and transform the way people perceive and interact with their environment. Furthermore, the proliferation of sustainable technologies and renewable energy sources promises to reshape global economies and mitigate ecological challenges. From advancements in quantum computing to the advent of space exploration, the possibilities are boundless. However, the ethical implications and societal adaptability must remain at the forefront of these developments. By critically examining historical breakthroughs and leveraging interdisciplinary insights, we can illuminate the path forward, ensuring that our technological evolution aligns with our collective values and aspirations

Integrative Technologies: Bridging Old and New

Integrative technologies represent the synergistic fusion of ancient and modern technological principles, fostering a harmonious convergence of tradition and innovation. This approach enables the preservation of time-honored techniques while integrating contemporary advancements, thereby bridging the gap between antiquity and the present. By leveraging Integrative Technologies, we can harness the cumulative wisdom of bygone eras and infuse it with the cutting-edge capabilities of today's rapidly evolving world.

At the core of Integrative Technologies lies the concept of amalgamating historic knowledge with progressive methodologies, resulting in bespoke solutions that address both historical challenges and present-day imperatives. Whether it involves leveraging ancient construction techniques to inform sustainable urban development or revitalizing traditional agricultural practices through modern automation and precision farming, the potential applications are multifaceted and boundless.

Moreover, Integrative Technologies serve as catalysts for cultural preservation, ensuring that ancient crafts, arts, and sciences are not consigned to the annals of history but are seamlessly woven into contemporary contexts. By embracing this ethos, societies can safeguard their cultural heritage while propelling themselves towards inclusive, forward-thinking approaches.

The interplay between historic mastery and modern ingenuity also offers unparalleled opportunities for cross-disciplinary collaborations. Whether it pertains to archaeological restorations informed by state-of-the-art 3D scanning and modeling or bioinspired design drawing from ancient biological principles, the synergy of disciplines results in innovative breakthroughs that transcend conventional boundaries.

Furthermore, the adoption of Integrative Technologies fosters an ecosystem where sustainability and efficiency converge, driving progress toward a more balanced and responsible future. Coupled with a conscientious reverence for environmental stewardship, this paradigm unlocks avenues for creative problem-solving grounded in the wisdom of antiquity and fortified by contemporary scientific and engineering prowess.

In essence, Integrative Technologies stand as conduits for a symbiotic relationship between the past and the future, offering transformative prospects across industries, cultures, and generations. Through their application, we honor the legacies of our forebears while charting a course towards a technologically enriched, socially inclusive, and environmentally sustainable global landscape.

Reflections on Sustainability and Resilience

In the wake of delving deep into the multi-faceted implications of ancient technologies, it becomes imperative to reflect on sustainability and resilience. As we examine the enduring legacies of civilizations that harnessed innovation millennia ago, we are compelled to scrutinize the sustainable practices embedded in their technological advancements. This introspection not only sheds light on how our predecessors cultivated a harmonious relationship with their environment, but also prompts us to reevaluate our current approaches towards resource management and conservation. By appreciating the resourcefulness of ancient civilizations, we gain valuable insights into developing sustainable technologies and fostering resilience in the face of environmental challenges. Moreover, this process inspires us to draw parallels between historical and contemporary endeavors, pinpointing opportunities to integrate traditional knowledge with modern solutions. The enduring legacy of sustainable practices from antiquity serves as a cornerstone for forging a more ecologically balanced future, ensuring that our innovations today align with the principles of sustainability and resilience. Embracing the wisdom of the past empowers us to create resilient systems and infrastructures that can withstand the test of time, mitigating the impact of natural calamities and human-induced disruptions. Moreover, by acknowledging the historical significance of sustainability and resilience, we are better equipped to navigate the complexities of a rapidly evolving technological landscape, where long-term stability is contingent on our ability to incorporate sustainable practices into the fabric of our advancements. This reflective exploration not only serves as a tribute to the resilience of ancient civilizations but also propels us towards a future where sustainability is non-negotiable, and where resilience is ingrained in the very essence of progress.

The Role of Education in Preserving Ancient Knowledge

Education plays a crucial role in the preservation and dissemination of ancient knowledge. As we reflect on the technological marvels of ancient civilizations, it becomes evident that the transmission of this knowledge is essential for its continuity and relevance in contemporary society. Through formal and informal educational platforms, the wisdom of our predecessors can be upheld and integrated into our modern understanding of technology and innovation.

Preserving ancient knowledge through education involves interdisciplinary approaches that merge historical, cultural, and technological studies. By incorporating ancient technological advancements into curricula, students gain a deeper appreciation for the ingenuity of past civilizations and an understanding of the foundations upon which our current technologies are built.

Furthermore, education provides a platform for interpreting and contextualizing ancient knowledge within the socio-cultural landscapes of diverse civilizations. By fostering an appreciation for ancient technology, educational institutions contribute to the preservation of these legacies, ensuring that they are not lost amidst the rapid advancement of modern innovations.

In addition to academic settings, informal educational initiatives such as museums, public lectures, and digital archives play a pivotal role in preserving ancient knowledge. These platforms engage a wide audience, fostering curiosity and appreciation for ancient technologies. By making this knowledge accessible to the public, these initiatives ensure that the legacies of ancient civilizations continue to inspire and inform present and future generations.

Moreover, integrating ancient knowledge into educational frameworks promotes critical thinking and encourages innovative problem-solving. Students exposed to the achievements and challenges faced by ancient technologists can draw valuable lessons that are pertinent to contemporary issues. By examining the societal impact of early inventions and considering the ethical implications of ancient technologies, learners develop a comprehensive understanding of how history informs present-day practices.

Ultimately, the role of education in preserving ancient knowledge extends beyond imparting factual information; it cultivates an enduring respect for the legacy of technological innovation. By weaving ancient knowledge into educational narratives, we equip future generations with the insights necessary to navigate the complexities of evolving technologies and foster a sustainable relationship with the past.

Concluding Thoughts: Where Do We Go from Here?

The preservation and dissemination of ancient knowledge through education are crucial steps in ensuring that the wisdom of our forebears continues to enrich and inspire contemporary and future generations. By integrating the teachings of ancient technology into modern educational curricula, we can empower individuals with a deeper understanding of the historical interconnectedness of innovation and the enduring relevance of ancient wisdom. This holistic approach to education not only fosters a sense of stewardship towards preserving ancient knowledge but also provides a wellspring of inspiration for interdisciplinary thinking and problem-solving. By encouraging students to critically engage with the achievements and challenges faced by ancient civilizations, we can cultivate a mindset that values resilience, sustainable practices, and ethical considerations in the pursuit of technological advancement. Additionally, promoting the study of ancient technology can help bridge the gap between traditional and STEM

(Science, Technology, Engineering, and Mathematics) disciplines, fostering a more comprehensive approach to contemporary technological challenges. As we navigate an increasingly complex and interconnected world, it is vital to acknowledge the profound lessons embedded within ancient technologies and the enduring legacies they carry. Embracing the insights from our historical predecessors allows us to envision a future shaped by a harmonious fusion of ancient wisdom and modern innovation. By cultivating a culture that honors the contributions of the past while embracing the possibilities of the future, we can pave the way for a renaissance of ancient technologies in a rapidly evolving world. Through collaborative research, knowledge exchange, and interdisciplinary collaboration, we can unlock the potential for revitalizing ancient technologies, infusing them with contemporary advancements to address pressing global issues such as environmental sustainability, cultural preservation, and societal well-being. The journey forward revolves around harnessing the collective knowledge and creativity of diverse communities, transcending disciplinary boundaries, and paying homage to the enduring legacy of ancient technologies.

19

The Future of Ancient Technology

Prospective Paradigms: Envisioning Tomorrow

In envisioning the future of technology, one must consider the remarkable potential lying at the intersection of ancient techniques and modern advancements. By extrapolating ancient wisdom and combining it with cutting-edge innovations, we can unlock a wealth of transformative possibilities. The concept of rejuvenating ancient technologies has sparked significant interest in various fields, including sustainability, engineering, and archaeology. Through a multidisciplinary approach, researchers and visionaries are exploring how age-old solutions to societal challenges can be reimagined and integrated with contemporary knowledge.

An essential aspect of this paradigm revolves around leveraging the sustainable practices of ancient civilizations to build a resilient, environmentally conscious future. The fusion of ancient agricultural methods with modern precision agriculture, for instance, holds immense promise in ensuring food security while minimizing ecological impact. By studying pre-industrial cultivation strategies, such as terracing and crop diversification, we gain insights into sustainable land use that can inform futuristic approaches to farming and land management.

Moreover, delving into ancient resource management techniques offers valuable lessons for contemporary endeavors aimed at responsible extraction and utilization of natural resources. Drawing from historical mining methods, which often emphasized a harmonious relationship with the environment, researchers are developing innovative mining practices that prioritize ecological preservation and long-term viability. Such endeavors showcase the potential for ancient technologies to guide us toward a more sustainable and balanced relationship with the planet.

The integration of ancient medical knowledge with modern healthcare presents another compelling avenue for future technological advancements. Comprehensive examinations of traditional healing practices not only serve to preserve cultural heritage but also yield novel insights that challenge conventional medical perspectives. By combining ancient herbal remedies and holistic therapies with advances in biotechnology and pharmacology, there exists an opportunity to develop personalized, nature-inspired treatments that align with individual wellness needs.

Furthermore, the revival of archaic craftsmanship and construction methods stands to revolutionize contemporary architecture and design. By drawing from the architectural prowess of past civilizations – encompassing structural durability, climate responsiveness, and aesthetic allure – architects and engineers are reimagining modern urban spaces with a nod to timeless design principles. Integrating ancestral construction techniques with state-of-the-art materials and methodologies promises to enhance the resilience and sustainability of future cities.

In essence, foreseeing the future of ancient technology entails a meticulous exploration of historical ingenuity integrated with forward-thinking innovation. This harmonious union allows us to anticipate transformative advancements across diverse sectors, offering glimpses into a future shaped by the enduring wisdom of antiquity.

Technologies Reborn: Potentials and Pitfalls

The resurrection of ancient technologies presents a realm of unprecedented potential and unforeseen pitfalls. As modern society rekindles its interest in the wisdom of our ancestors, it is essential to recognize the myriad opportunities and challenges that accompany this resurgence. By harnessing the innovations of antiquity, we embark on a journey that promises transformative breakthroughs across various sectors, from sustainable urban development to medical advancements.

Revisiting age-old methodologies compels us to confront the inherent risk of misinformation and misinterpretation. Without rigorous scrutiny and validation, there exists a looming danger of inadvertently propagating inaccuracies and deceptions. Moreover, the incomplete understanding of historical contexts may lead to misguided applications, breeding unintended consequences that could impede progress and innovation.

Conversely, the potential for growth and evolution through the reincarnation of ancient technologies cannot be overstated. By marrying historical insights with contemporary scientific knowledge, we stand to unlock a trove of untapped possibilities. From agricultural practices that prioritize ecological balance to architectural designs informed by time-honored engineering principles, the resurrection of ancient technologies nurtures a landscape rich with promise.

However, the integration of revived practices into modern frameworks demands a nuanced approach. Delicate ethical and cultural considerations must be carefully navigated to ensure that these resurfaced modalities are implemented with respect and sensitivity. Additionally, identifying the boundaries between homage to tradition and exploitation of heritage is imperative in avoiding the commercialization and commodification of ancient wisdom.

Finally, as we venture into uncharted territory, the interplay between technological revival and contemporary paradigms necessitates continuous vigilance and critical evaluation. The potentials of reviving ancient technologies are vast, and the accompanying pitfalls are equally formidable. Only through meticulous discernment, collaboration, and commitment to ethical stewardship can we hope to tap into the boundless reservoir of benefits while safeguarding against the perils that lurk in the shadows.

Synergies with Modern Science

The intersection of ancient technology with modern science presents a compelling landscape for exploration and innovation. By bridging the wisdom of antiquity with contemporary scientific methodologies, we open the door to unprecedented opportunities in various fields. One of the most fascinating aspects is the potential synergy between ancient engineering marvels and cutting-edge scientific disciplines such as materials science, nanotechnology, and biotechnology.

In the realm of materials science, the advanced knowledge embedded in ancient artifacts and structures offers valuable insights into the development and application of novel materials with exceptional properties. For instance, the durability and resilience of ancient concrete used in Roman architecture have stimulated research into more sustainable and durable construction materials. By studying these ancient techniques, scientists and engineers can develop innovative materials that are not only eco-friendly but also economically viable.

Moreover, the integration of ancient wisdom with nanotechnology holds promise for revolutionary advancements. The intricate craftsmanship displayed in ancient artifacts points to an understanding of nanoscale phenomena, suggesting that our predecessors might have manipulated materials at the molecular level. By leveraging this knowledge, modern scientists can explore new frontiers in nanofabrication, enabling the creation of high-performance materials, microdevices, and medical technologies with unprecedented precision and functionality.

Biotechnology also stands to benefit from the convergence of ancient knowledge and modern science. The rich history of herbal medicine, botanical remedies, and traditional healing practices across diverse cultures provides a treasure trove of natural compounds and biological insights. By scrutinizing these ancient remedies through the lens of contemporary biochemistry and pharmacology, researchers can uncover novel therapeutic agents, develop new drug formulations, and gain a deeper understanding of human health and wellness.

Furthermore, the fusion of ancient technology and modern science extends beyond material and biological sciences. It encompasses interdisciplinary collaborations in environmental sustainability, aerospace engineering, and information technology. By delving into the principles and achievements of our ancestors, scientists and innovators can devise pioneering solutions to pressing global challenges, propel space exploration endeavors, and engineer sophisticated computational systems inspired by ancient computational devices.

Embracing the synergies with modern science allows us to tap into a vast reservoir of untapped knowledge and ingenuity. As we navigate this convergence, it is crucial to approach this fusion with respect for cultural heritage, ethical considerations, and a commitment to sustainable progress. By harmonizing the timeless wisdom of ancient technology with the dynamism of modern science, we pave the way for monumental discoveries, paradigm-shifting innovations, and a profound appreciation for the enduring legacy of human achievement.

Sustainable Development Through Ancient Insights

The quest for sustainable development has become an urgent global priority, with increasing concern over environmental degradation and resource depletion. As modern societies grapple with these challenges, there is a growing realization that ancient civilizations had unique insights and practices that could offer valuable lessons for sustainable living. By examining the environmental ethos of ancient cultures, we can glean wisdom that may be instrumental in guiding our contemporary pursuit of sustainability. From indigenous communities' deep reverence for nature to ancient agricultural techniques that promoted soil fertility and biodiversity, there is much to be learned from the sustainable practices of our ancestors. The preservation of ancient ecosystems and wildlife sanctuaries by early civilizations indicates an intrinsic understanding of the delicate balance between human activities and the natural environment. Drawing inspiration from these ancient conservation efforts, modern societies can develop strategies to preserve and protect precious natural habitats. Furthermore, ancient approaches to waste management and recycling demonstrate innovative solutions that echo current calls for circular economies and reduced ecological footprints. Through a comprehensive study of historical sustainable practices, we stand to gain profound insights into harmonizing human activities with the environment. Integrating ancient wisdom with cutting-edge technologies and practices offers promising avenues for achieving sustainable development goals. By leveraging the best of both worlds, we can co-create a future where ecological harmony and human advancement are mutually reinforcing. Embracing the tenets of sustainable development through ancient insights holds the promise of a more balanced and resilient society, fulfilling the needs of the present without compromising the welfare of future generations.

Innovative Approaches in Reviving Ancient Practices

Reviving ancient practices requires innovative approaches that integrate traditional wisdom with modern technologies and methodologies. Through collaborative efforts between historians, archaeologists, anthropologists, and technology experts, various groundbreaking initiatives have emerged to explore the potential of ancient practices in addressing contemporary challenges. One such approach involves leveraging digital reconstruction and simulation techniques to recreate and understand ancient technologies and manufacturing processes. By scrutinizing and replicating age-old methods, researchers aim to unlock valuable insights into sustainability, craftsmanship, and resource efficiency inherent in these practices. Another innovative approach focuses on revitalizing ancient agricultural techniques, such as terrace farming and aqueduct systems, to promote sustainable food production and water management. By incorporating cutting-edge irrigation systems and precision agriculture technologies, these age-old practices can play a crucial role in mitigating modern environmental concerns. Furthermore, the interdisciplinary integration of traditional craftsmanship with modern design and engineering principles has opened avenues for reviving ancient artisanal techniques in crafting durable and aesthetically captivating products. This fusion not only preserves cultural heritage but also yields novel solutions for contemporary design challenges. Additionally, modern medical research is exploring the efficacy of ancient herbal remedies and healing practices, aiming to integrate them into evidence-based healthcare systems. Through rigorous scientific validation, traditional medicines are increasingly gaining recognition for their therapeutic potential, offering new pathways for holistic wellness. Moreover, innovative conservation strategies employ advanced materials science and preservation technologies to safeguard ancient structures, artifacts, and documents. These endeavors ensure the longevity of cultural heritage while facilitating in-depth studies of ancient knowledge and skills. Embracing innovative approaches in reviving ancient practices requires due diligence in understanding the historical, cultural, and social contexts surrounding these traditions. Additionally, it demands respectful engagement with indigenous communities and stakeholders to uphold ethical considerations, respect intellectual property rights, and ensure equitable sharing of benefits. Collaborative partnerships and inclusive decision-making processes are pivotal in cultivating mutual trust and fostering sustainable development through the revival of ancient practices. As we chart this uncharted territory, navigating the intersection of tradition and innovation, innovative approaches offer promising avenues for preserving and harnessing the timeless wisdom encapsulated in ancient practices.

Ethical Considerations and Responsibilities

In the endeavor to revive ancient technologies, it is imperative to address the ethical considerations and responsibilities associated with this pursuit. As we explore the potential applications of age-old practices in the modern world, it becomes crucial to uphold ethical principles that preserve cultural integrity, respect indigenous knowledge, and ensure equitable collaboration. The responsibility lies not only in using ancient technologies for societal development but also in safeguarding the rights and traditions of the communities from which these technologies originated. Ethical considerations extend to issues such as intellectual property rights, fair compensation, and informed consent when engaging with traditional knowledge holders. Furthermore, the ethical implications of resurrecting ancient technologies also encompass environmental sustainability and the impact on local ecosystems. It is essential to assess the ecological effects of implementing ancient practices on a larger scale, considering their historical context and current environmental dynamics. This includes evaluating the potential consequences of widespread adoption of ancient techniques on biodiversity, land use, and natural resources. Moreover, the ethical framework must also address the potential social and economic repercussions of integrating ancient technologies into contemporary systems. This involves analyzing the potential disruption to existing industries, employment patterns, and socioeconomic structures. Caution must be exercised to mitigate any adverse effects on vulnerable communities while leveraging ancient technologies for progress. Additionally, ethical responsibilities call for establishing transparent governance mechanisms and regulatory frameworks to prevent exploitation, appropriation, or misappropriation of ancient wisdom and heritage. Respectful engagement with indigenous and local communities forms the cornerstone of ethical practices in advancing ancient technologies. Collaborative partnerships should prioritize mutual respect, reciprocity, and capacity building within these communities, acknowledging their intrinsic expertise and wisdom. Beyond recognizing the value of traditional knowledge, ethical responsibilities encompass promoting inclusive and culturally sensitive approaches that empower these communities while fostering sustainable development. It is imperative to engage in continuous dialogue, honoring the voices and agency of these communities throughout the process of reviving ancient technologies.

Recognizing the complex web of ethical considerations and responsibilities is paramount in guiding the responsible revival of ancient technologies. By upholding ethical standards that align with principles of justice, equity, and respect, we can navigate the evolving landscape of technological revitalization with mindfulness and integrity.

Impact on Global Economies and Policies

The integration of ancient technology into modern practices has the potential to significantly impact global economies and policies. As societies rediscover and incorporate ancient innovations, there is a likelihood of economic shifts due to changes in production processes, resource management, and market demand. Additionally, the emergence of new industries and business opportunities associated with the revival of ancient technologies may lead to the reshaping of local and international markets. Moreover, the adoption of sustainable practices derived from ancient wisdom could influence policies related to environmental conservation, energy utilization, and international trade agreements. These changes may necessitate the development of regulatory frameworks and standards to govern the application of ancient technologies in various economic sectors. From an investment perspective, the resurgence of ancient technologies presents new opportunities for capital allocation and the diversification of portfolios, particularly in emerging markets and sectors such as renewable energy, traditional medicine, and eco-friendly manufacturing processes. While these developments offer prospects for economic growth and innovation, they also pose challenges in terms of competition, intellectual property rights, and ethical business practices. Therefore, the impact on global economies and policies warrants careful consideration and strategic planning to ensure sustainable and equitable growth in the era of renewed ancient technological advancements.

Educational Implications and Knowledge Dissemination

The exploration of ancient technology not only offers profound insights into our history, but also presents a unique opportunity to reshape educational paradigms. Integrating the study of ancient technologies into educational curricula can enrich and diversify learning experiences for students at all levels. The implications are vast, touching on various disciplines such as history, archaeology, anthropology, engineering, and more.

At its core, the study of ancient technology promotes critical thinking and problem-solving skills. By delving into the innovations of past civilizations, students are encouraged to analyze, interpret, and engage with complex concepts that have relevance to contemporary society. Through hands-on explorations and practical experiments, learners can gain a deeper understanding of the ingenuity that drove technological advancements in ancient times.

Furthermore, knowledge dissemination plays a pivotal role in ensuring the preservation and continued study of ancient technologies. Academic institutions, museums, and cultural organizations have a responsibility to convey the significance of these discoveries to wider audiences. Implementing outreach programs, workshops, and public exhibitions can foster public interest and awareness, thereby contributing to the appreciation and safeguarding of ancient technological legacies.

Incorporating these educational initiatives also cultivates a sense of global citizenship and interconnectedness. Students are exposed to diverse cultural perspectives and are prompted to consider the profound impact of ancient technologies on present-day societies. This approach fosters a respect for cultural heritage and fosters collaboration across geographical and disciplinary boundaries, promoting a broader understanding of the human experience.

The reimagining of educational frameworks to encompass ancient technology encourages interdisciplinary dialogue and collaboration among scholars, educators, and practitioners. It serves as a platform for fostering innovation and creativity while grounding contemporary advancements in historical context. As such, educational institutions play a vital role in shaping the narratives surrounding ancient technology, ensuring that the knowledge is upheld, critically examined, and perpetuated for future generations.

Technological Integration and Cultural Challenges

The integration of ancient technology into modern society presents a unique set of challenges, particularly in relation to cultural dynamics and societal norms. As we strive to incorporate ancient practices and knowledge into contemporary technological frameworks, it becomes imperative to navigate the intricate web of cultural nuances and sensitivities. One of the primary considerations in this endeavor is the potential clash between traditional cultural beliefs and the adoption of advanced technologies rooted in ancient wisdom. Moreover, there exists a delicate balance in ensuring that such integration respects and preserves the cultural heritage from which these technologies originate. It is essential to approach this integration with a deep understanding and appreciation of diverse cultural traditions, as well as a profound respect for indigenous knowledge. At the core of this challenge lies the need to facilitate a symbiotic relationship between ancient technology and cultural heritage, fostering mutual enrichment without overshadowing or diluting the intrinsic value of either aspect. The process of technological integration also demands thoughtful examination of power dynamics and societal implications. It is crucial to acknowledge and address any disparities in access to this integrated technology, ensuring that all communities have equal opportunities to benefit from its advancements. This necessitates extensive collaboration with local authorities, cultural custodians, and community representatives to establish inclusive frameworks for the dissemination and application of ancient technological solutions. Furthermore, the ethical considerations surrounding this integration cannot be overlooked.

As we embark on this journey, ethical guidelines must be established to govern the responsible adaptation and use of ancient technology within modern contexts. The careful consideration of these ethical parameters is vital in minimizing potential harm and maximizing the positive impact of technological integration on societies. Ultimately, successfully addressing the cultural challenges associated with technological integration necessitates a holistic, collaborative approach that values and incorporates diverse perspectives. By fostering meaningful dialogue, engaging in close consultation with relevant stakeholders, and upholding a deep commitment to cultural sensitivity, we can pave the way for a harmonious integration of ancient technology within our modern world.

The Next Frontier: Uncharted Territories

In examining the future of ancient technology, one cannot overlook the concept of uncharted territories. The convergence of ancient wisdom and modern expertise paves the way for unexplored frontiers that hold tremendous potential. As we delve into these uncharted territories, it becomes evident that ancient technology forms the bedrock upon which novel innovations can thrive. In this domain, the fusion of historical insight and contemporary advancements presents boundless opportunities to redefine the landscape of technological progress.

At the forefront of uncharted territories lies the realm of interdisciplinary collaboration. By breaking traditional boundaries and fostering synergies across various fields of knowledge, we can unlock new vistas of possibility. Furthermore, the exploration of uncharted territories in ancient technology necessitates a keen focus on sustainability and ethical practices. Understanding and embracing the ecological, social, and cultural impacts of technological advancements is crucial in charting a responsible course into the future.

Moreover, as we venture into these unexplored domains, it is imperative to acknowledge the inevitable challenges that lie ahead. The inherent complexities associated with integrating ancient technologies with modern frameworks necessitate meticulous planning and thorough understanding. This journey into uncharted territories calls for a harmonious balance between preservation and evolution, ensuring that the essence of ancient practices is upheld while accommodating the demands of contemporary society.

Amidst these endeavors, educational initiatives play a pivotal role in navigating the uncharted territories of ancient technology. Implementing comprehensive programs that elucidate the historical relevance of ancient practices and their potential applications in modern contexts is fundamental to shaping a well-informed and empowered global community. Additionally, fostering open dialogue and international cooperation is indispensable in navigating the uncharted territories, fostering a collective vision for the future.

As we embark on the pursuit of uncharted territories in ancient technology, it is imperative to recognize that this journey is characterized by both discovery and responsibility. The exploration of untapped potential demands a balanced approach, one that respects the wisdom of the past while embracing the innovation of the present. Ultimately, the uncharted territories harbor a wealth of promise, offering a compelling narrative of continuous progress and enlightenment.

20

Renaissance of Ancient Technology

Revival and Relevance: Introduction to Ancient Technological Resurgence

The resurgence of interest in ancient technologies is a testament to the enduring allure of human innovation across millennia. As contemporary society continues to grapple with increasingly complex challenges, the re-evaluation of ancient technological practices offers a unique avenue for inspiration and problem-solving. The scope of application for these revived techniques spans a broad spectrum, encompassing fields as diverse as sustainable architecture, renewable energy, agricultural practices, and even medical treatments. Furthermore, the re-emerging technologies from antiquity provide valuable insights into cultural heritage and historical continuity, fostering a deeper appreciation for the achievements of our predecessors. By delving into the revival and relevance of ancient technologies, we embark on a journey that not only honors the ingenuity of the past but also paves the way for innovative solutions in the present and future.

Historical Context: Re-emerging Technologies from Antiquity

As civilization progresses, the value of ancient technologies is increasingly recognized. Historical context plays a pivotal role in understanding the re-emergence of these technologies from antiquity. The revival of ancient technologies provides an opportunity for modern society to bridge the gap between historical wisdom and contemporary innovation. By delving into the historical context, we gain insights into the diverse cultural and technological contributions made by ancient civilizations. The resurgence of ancient technologies reflects not only a desire to preserve heritage but also an eagerness to unearth age-old wisdom that can enrich modern practices.

The historical context encompasses a broad spectrum of disciplines, including archaeology, anthropology, and cultural studies. It involves careful examination of archaeological findings, ancient texts, and artistic representations to comprehend the technological prowess of bygone eras. By scrutinizing historical records and artifacts, researchers can decipher the methodologies and applications of ancient technologies, shedding light on their significance in the development of human civilization.

Furthermore, understanding the historical context offers a nuanced perspective on the social, economic, and environmental factors that influenced the evolution and dissemination of ancient technologies. Ancient societies' adaptive strategies, resource utilization, and technological innovations provide valuable lessons applicable to contemporary challenges. From the irrigation techniques of ancient Mesopotamia to the architectural marvels of the Incas, each era presents a tapestry of technological ingenuity waiting to be unveiled and integrated into the present.

Moreover, the historical context prompts critical analysis of the transmission and preservation of ancient knowledge through generations. It reveals the interconnectivity of civilizations and their contributions to a shared repository of wisdom. Through the exploration of historical context, a narrative emerges, highlighting the resilience and adaptability of ancient technologies as they were assimilated, transcended, and revitalized by successive societies. This historical continuum portrays the enduring relevance of ancient technologies and their potential to address modern challenges with unconventional solutions.

In summary, delving into the historical context reveals the intrinsic value of ancient technologies and their enduring legacy. It provides a foundation for appreciating the sophisticated systems developed by early cultures and underscores the significance of integrating ancient wisdom into contemporary technological endeavors.

Methodologies in Recovering Lost Technologies

The pursuit of reviving lost technologies from ancient civilizations is a complex and multidisciplinary endeavor that requires a strategic approach and a deep understanding of historical, archaeological, and technological domains. To commence the process of recovering lost technologies, extensive research and investigation into historical texts, artifacts, and archaeological findings are essential. This involves collaboration between historians, archaeologists, anthropologists, and technologists to decipher ancient inscriptions, analyze artifacts, and reconstruct the operational principles and mechanisms of these technologies. The utilization of advanced imaging techniques such as 3D scanning, X-ray spectroscopy, and computational modeling play pivotal roles in understanding the intricacies of these ancient marvels. Furthermore, the integration of material science and metallurgy aids in identifying and replicating the materials used in the construction of these technologies, providing valuable insights into ancient craftsmanship.Engaging with indigenous communities and custodians of traditional knowledge is also crucial in uncovering lost technologies, as these communities may possess oral traditions or practical knowledge pertaining to ancient techniques and practices. Ethnographic studies and fieldwork enable researchers to understand the contextual nuances and socio-cultural significance associated with these technologies, contributing to a holistic recovery process. Moreover, interdisciplinary workshops, symposiums, and collaborative projects facilitate knowledge exchange and skill enhancement among experts specializing in diverse fields, fostering a collective effort towards the recuperation of ancient technologies. The process of recovering lost technologies does not merely involve replication but also necessitates adaptation and innovation. Integrating recovered ancient technologies with contemporary advancements involves a careful balance between preserving the authenticity of the original designs and incorporating modern scientific understanding and engineering principles. Utilizing digital simulations, rapid prototyping, and experimental reconstructions enables the validation and refinement of recovered technologies, ensuring their viability and functionality in present contexts. Additionally, engaging with contemporary artisans and expert craftsmen contributes to the practical reconstruction of ancient technologies, combining traditional craftsmanship with modern precision. As the endeavor to recover lost technologies progresses, it is imperative to consider the ethical dimensions surrounding heritage preservation and global cultural diversity. Respecting the intellectual property rights of indigenous communities and acknowledging the cultural significance of these technologies are integral aspects of responsible recovery efforts. By adhering to ethical guidelines and international regulations, the revival of lost technologies can be harmoniously integrated into educational, cultural, and scientific domains, enriching our collective understanding of human ingenuity across millennia. In essence, the methodologies employed in recovering lost technologies embody a fusion of academic rigor, technological innovation, cross-disciplinary collaboration, and cultural stewardship, paving the way for the renaissance of ancient wisdom in contemporary times.

Technological Integration: Blending Old with New

The integration of ancient technology with modern innovations marks a pivotal moment in the evolution of human progress. This process involves the reimagining and adaptation of age-old techniques to suit contemporary needs and requirements. It requires a delicate balance of preserving the essence of ancient wisdom while leveraging the advancements of the present era.

Modern science and engineering have provided us with valuable insights into the fundamental principles behind ancient technologies, enabling us to comprehend their intricacies and functionalities. By extracting these core concepts, we can identify opportunities for integration onto our existing technological frameworks. This fusion unravels new potentials and propels us towards uncharted territories, combining the wisdom of the past with the resources of the present.

The seamless blending of old and new technologies often involves interdisciplinary collaboration, where experts from diverse fields come together to create synergies that transcend the limitations of isolated domains. For instance, material scientists may examine the composition of ancient artifacts, seeking inspiration for the development of novel materials with enhanced properties. Similarly, engineers might study the structural design of ancient constructions to refine architectural methodologies or develop innovative building materials.

Moreover, the integration of ancient technology goes beyond tangible artifacts; it encompasses knowledge systems, philosophies, and societal practices that offer profound insights into sustainable living, resource management, and ecological balance. Lessons learned from ancient civilizations regarding harmonious coexistence with nature serve as beacons guiding contemporary efforts to nurture a symbiotic relationship between humanity and the environment.

By amalgamating ancient technologies with modern practices, we not only honor the legacy of our ancestors but also embark on a journey of discovery and innovation. It stimulates a collective awakening, fostering an appreciation for historical ingenuity while invigorating our pursuit of technological excellence. The interplay between tradition and transformation provides a fertile ground for creativity, sparking ingenious solutions to pressing challenges and inspiring novel approaches to scientific and engineering endeavors.

Case Studies: Successful Modern Implementations of Ancient Tech

In today's world, the integration of ancient technology into modern applications has showcased remarkable feats of innovation and problem-solving. Through meticulous research, archaeologists and scientists have unearthed ancient tools, techniques, and knowledge that have been successfully adapted to address contemporary challenges. One exemplary case study involves the revitalization of ancient irrigation systems used by civilizations such as the Indus Valley and the Hohokam culture in the American Southwest. By studying their sophisticated networks and hydraulic engineering methods, modern engineers have been able to enhance agricultural sustainability and water management in arid regions, significantly impacting local communities' livelihoods and food security. Another compelling example pertains to the utilization of ancient herbal remedies and medicinal practices from diverse cultures, including traditional Chinese medicine and Ayurveda, in developing alternative and complementary healthcare solutions. By integrating age-old wisdom with modern scientific validation, researchers and practitioners have enriched therapeutic options and expanded the understanding of holistic wellness. Moreover, breakthroughs in material science have seen the resurgence of ancient construction methods, such as using sustainable and durable materials like rammed earth and cob for eco-friendly architecture and infrastructure projects. These successful implementations demonstrate the value of learning from our ancestors and leveraging their timeless wisdom to tackle pressing issues in today's society. As we continue to delve into the wealth of ancient knowledge, further opportunities for innovation and sustainable development are poised to arise, showcasing the enduring relevance and applicability of ancient technology in reshaping our contemporary world.

Challenges in Contemporary Resurrection of Ancient Techniques

The contemporary resurrection of ancient techniques presents a myriad of challenges that necessitate careful consideration and strategic approaches. While the potential benefits are tantalizing, the path to integrating age-old technologies into modern contexts is fraught with complexities. One major challenge lies in the lack of comprehensive documentation and understanding of ancient methodologies. Many of these techniques were passed down orally through generations or were lost to the ravages of time, leaving fragmented clues for contemporary researchers and innovators to piece together. This dearth of knowledge poses a significant obstacle, as it requires interdisciplinary collaboration and an innovative mindset to fill in the gaps and decipher the cryptic remnants of bygone technologies.

Furthermore, the juxtaposition of ancient methodologies with contemporary standards brings forth challenges in scalability and reproducibility. The integral question emerges: How can these ancient techniques be effectively adapted and implemented on a large scale without compromising their authenticity and efficacy? Striking a delicate balance between preserving the integrity of ancient practices while ensuring compatibility with modern demands necessitates meticulous planning and empirical testing. Moreover, the need to navigate regulatory frameworks and intellectual property rights adds another layer of complexity, requiring conscientious navigation to avoid conflicts and adhere to ethical standards.

Amidst these challenges, the ethical implications of resurrecting ancient technologies manifest prominently. Considerations regarding cultural appropriation, respect for indigenous knowledge, and the potential distortion of historical legacies must be carefully addressed. It is imperative to engage with relevant communities and stakeholders to ensure that the revival of ancient techniques is conducted with the utmost sensitivity and respect. Additionally, the preservation of heritage and the long-term impact on local societies and ecosystems cannot be overlooked. Achieving a harmonious coalescence between progress and preservation necessitates a nuanced approach rooted in mutual understanding and shared benefit.

Moreover, the limitations of ancient materials and tools present tangible hurdles in the contemporary revitalization of archaic technologies. Fostering an understanding of the inherent constraints and exploring innovative means of supplementing or enhancing these materials without diluting their historical essence poses a formidable challenge to researchers and practitioners alike. Furthermore, issues of knowledge transmission and intergenerational learning become apparent, as the loss of traditional expertise compounds the intricacies of reviving ancient techniques.

In navigating these challenges, it becomes evident that the resurrection of ancient technologies is a multifaceted endeavor that requires a holistic and inclusive approach. Proactive collaboration between diverse fields of expertise, adherence to ethical considerations, and a steadfast commitment to preserving the intrinsic value of ancient techniques are indispensable in overcoming the obstacles that accompany the contemporary resurgence of age-old technologies.

Ethical Considerations and the Preservation of Heritage

Preserving and reviving ancient technologies brings with it a host of ethical considerations that must be carefully navigated to ensure the preservation of heritage. The resurgence of ancient techniques often involves knowledge that has been passed down through generations or retrieved from historical records, indicating a deep-seated connection to cultural identity and legacy. It becomes imperative for contemporary society to embrace ethical practices when engaging with these technologies. One of the key ethical considerations is the respect for indigenous knowledge and intellectual property rights. Indigenous communities may hold traditional knowledge of ancient technologies, and it is essential to recognize their ownership and rights over this knowledge. Collaboration and partnerships with indigenous groups can ensure that the revival of ancient technologies is conducted with respect and acknowledgement of their cultural significance. Moreover, the preservation of heritage requires a delicate balance between technological progress and the safeguarding of cultural integrity. While modern advancements offer opportunities to revive and adapt ancient technologies, it is crucial to maintain the authenticity and original practices embedded in these inventions. Ethical frameworks should be established to guide the responsible integration of ancient techniques into contemporary contexts, considering the cultural, social, and environmental implications. Another critical aspect pertains to sustainability and environmental impacts. As ancient technologies are resurrected, it is vital to assess their ecological footprint and implement sustainable practices. Responsible utilization of natural resources and adherence to environmentally friendly processes are fundamental in preserving the equilibrium between innovation and conservation. Additionally, ethical considerations extend to the impact on local communities and societies. The revival of ancient technologies has the potential to influence livelihoods, economic systems, and social structures. Ensuring that these advancements provide equitable benefits to the communities involved and do not disrupt societal harmony is an ethical imperative. This requires comprehensive engagement with stakeholders, transparent communication, and equitable distribution of the fruits of technological renaissance. Ultimately, the preservation of heritage through the renaissance of ancient technology necessitates a conscientious approach that honors diverse cultures, respects traditional knowledge holders, promotes sustainability, and fosters positive societal impacts. By upholding ethical standards throughout the process, the legacy of ancient technologies can be upheld while contributing to progressive advancements in a responsible and inclusive manner.

Impact on Modern Science and Engineering

The renaissance of ancient technology has had a profound impact on modern science and engineering, sparking a resurgence of interest in traditional methods and materials. By studying the innovations of our forebears, contemporary scientists and engineers have gained unique insights into sustainable practices, eco-friendly designs, and innovative solutions to complex problems. The application of ancient technologies in modern contexts has not only broadened our understanding of historical achievements but has also led to groundbreaking advancements in various fields. One notable area of impact is in the realm of materials science. Through the exploration of age-old construction techniques and material compositions, researchers have discovered new ways to develop resilient and long-lasting materials, with properties that outperform many contemporary counterparts. Additionally, the integration of ancient wisdom into modern engineering has revolutionized the approach to sustainable infrastructure, leading to the creation of more durable and environmentally friendly buildings, bridges, and urban planning projects. Moreover, the influence of ancient technology has extended to the field of bioengineering, where the study of ancient medical practices and anatomical knowledge has inspired innovative approaches to healthcare and biotechnology. By applying ancient medicinal practices alongside modern scientific methods, researchers have uncovered valuable remedies and treatment modalities, tapping into a wellspring of natural and historically proven therapies. Furthermore, the impact of ancient technology on modern science can be observed in the realms of energy production and consumption. By revisiting traditional sources of energy, such as water wheels, windmills, and solar architecture, engineers have developed sustainable energy solutions that are both efficient and environmentally conscious, paving the way for a greener future. In addition, the study of ancient astronomical instruments and navigational tools has enhanced modern space exploration and geographical mapping, providing invaluable insights into celestial mechanics and Earth's geography. This synergy between past and present has not only expanded the horizons of scientific inquiry but has also fostered a deeper appreciation for the ingenuity of ancient civilizations, catalyzing a renaissance of learning and discovery that continues to shape the landscape of contemporary science and engineering.

Future Directions: Potential and Projections

As we stand at the cusp of a renaissance in ancient technology, it is imperative to explore the potential and projections for its future integration into contemporary society. The revival of ancient techniques presents a myriad of possibilities across various fields, including engineering, archaeology, art restoration, and cultural preservation. One key area of potential lies in the application of ancient construction methods to modern architecture and infrastructure. By studying and implementing the construction principles of ancient civilizations, we can create sustainable, durable, and aesthetically pleasing structures that harmonize with their natural surroundings. Additionally, the resurgence of ancient agricultural practices offers the potential to address contemporary sustainability challenges. Techniques such as terraced farming, aqueduct systems, and crop diversity, honed by ancient societies, may hold the key to ensuring food security and ecological balance in our rapidly evolving world. The fusion of ancient wisdom with cutting-edge technology also holds immense promise. By leveraging insights from ancient texts and artifacts, combined with advanced scientific methodologies, we can innovate in areas such as materials science, medicine, and energy production. Furthermore, exploring the potential of ancient technology in space exploration and off-world colonization represents a captivating frontier. The adaptability and resilience inherent in ancient techniques could prove invaluable in extraterrestrial endeavors. In projecting the future of ancient technology, it is crucial to emphasize the need for interdisciplinary collaboration, ethical considerations, and engagement with indigenous communities. While the potential benefits are vast, it is essential to ensure that the resurgence of ancient technology respects cultural heritage, fosters inclusivity, and promotes sustainable development. This necessitates close collaboration between historians, scientists, engineers, and local communities to ethically and responsibly integrate ancient techniques into contemporary practices. In conclusion, the future of ancient technology holds boundless potential to revolutionize diverse aspects of our lives. By embracing this renaissance and navigating its trajectory with foresight and integrity, we can unlock innovative solutions to pressing global challenges and honor the ingenuity of our ancestors.

Conclusion: Implications for Future Research and Development

As we look to the future, the renaissance of ancient technology presents profound implications for both research and development. The revival of age-old engineering marvels not only offers a window into the past but also paves the way for innovative possibilities in contemporary and future applications.

One significant implication lies in the realm of materials science and engineering. By studying and implementing ancient construction techniques, ranging from stone masonry to metalworking, researchers can gain insights into sustainable and durable materials that have withstood the test of time. This exploration can fuel innovations in modern construction and infrastructure projects, addressing challenges such as environmental impact and longevity.

Furthermore, delving into the methods and principles behind ancient technological achievements opens doors to novel approaches in various scientific disciplines. For instance, the intricate knowledge inherited from historical perspectives on medicine, astronomy, and mathematics can inspire new discoveries with relevance to modern healthcare, space exploration, and computational algorithms.

Moreover, the integration of ancient wisdom with contemporary technology has the potential to revolutionize fields like robotics, automation, and artificial intelligence. By studying the rudimentary yet sophisticated automata of antiquity, engineers may uncover unconventional design paradigms that optimize efficiency, adaptability, and human-machine interaction.

The implications are not merely technical; they extend to societal and cultural dimensions as well. In the backdrop of globalization and the whirlwind evolution of digital technologies, the resurgence of ancient practices fosters a balanced appreciation for tradition within an ever-transforming world. This cultural preservation and reinforcement of heritage through technological reawakening can engender a sense of pride, identity, and continuity, thereby enriching the fabric of society.

Looking ahead, future research endeavors must strive for meticulous documentation and preservation of the rediscovered ancient technologies, ensuring ethical and responsible implementation. This necessitates interdisciplinary collaborations among historians, archaeologists, engineers, and ethicists to navigate the nuanced landscapes of heritage conservation and technological advancement. Furthermore, fostering international cooperation and exchange of knowledge will be fundamental in harnessing the collective wisdom encapsulated in the legacy of ancient technology.

21

Exploring Ancient Technologies in Science Fiction

Ancient Technologies in Science Fiction

In exploring the fusion of ancient technologies and science fiction literature, it becomes evident that authors have long been fascinated by the prospect of integrating historical innovation with speculative future advancements. This intricate blend allows for a captivating narrative where antiquated mechanisms intertwine with futuristic possibilities, offering readers a thought-provoking journey through the corridors of time. The historical overview of ancient tech in sci-fi literature unveils the profound role played by ancient innovations in shaping fictional landscapes and technological frontiers within the genre. As writers delve into the depths of history to extract inspiration, they ingeniously weave ancient artifacts, inventions, and concepts into tales that transcend the boundaries of time and imagination.

Within this exploration, various science fiction subgenres offer diverse approaches to incorporating ancient technologies. Whether through steampunk's reimagining of Victorian-era inventions, or through space operas featuring lost civilizations with advanced knowledge, the breadth of literary expression is vast. Authors meticulously showcase how these age-old technologies are not mere anachronisms, but potent elements that enrich the fabric of their narratives, adding depth and complexity to their imagined worlds.On one hand, some stories depict ancient technologies as remnants of forgotten eras, holding mysterious powers waiting to be harnessed. On the other hand, certain works envision a parallel evolution where ancient knowledge thrives alongside futuristic marvels, blurring the lines between past and future.

Moreover, the integration of ancient technologies in science fiction invites contemplation on the implications of such amalgamations. Readers are compelled to ponder the nature of progress and innovation, prompting introspection on humanity's relationship with its historical legacy. This dynamic interplay sheds light on the cyclical nature of technological advancement, bridging past and future with profound philosophical resonance.Such rich thematic explorations establish ancient technologies as more than just plot devices, elevating them to pivotal motifs that drive the narrative and engage readers on intellectual and emotional levels. As we embark on this expedition through the annals of science fiction, the intersection of ancient and future technologies promises a captivating odyssey of speculative wonder and philosophical introspection.

Historical Overview of Ancient Tech in Sci-Fi Literature

Science fiction literature has long been fascinated with the incorporation of ancient technologies into its narratives, blending archaeological myths and speculative futures. The exploration of ancient tech in science fiction can be traced back to the early works of authors who delved into themes of lost civilizations and enigmatic artifacts. H.G. Wells, a pioneer of the science fiction genre, wove elements of ancient technologies and futuristic inventions in his seminal works such as 'The Time Machine' and 'The War of the Worlds.' These foundational texts set the stage for the subsequent integration of ancient technologies into the evolving landscape of science fiction. The mid-20th century marked a significant shift in the portrayal of ancient tech in sci-fi literature, with authors like Arthur C. Clarke exploring the concept of alien technologies that were indistinguishable from magic, blurring the line between ancient and extraterrestrial advancements. Isaac Asimov's exploration of robotics and artificial intelligence also contributed to the incorporation of ancient motifs into the futuristic narratives of science fiction. As the genre expanded, authors such as Frank Herbert and Ursula K. Le Guin further explored the convergence of ancient and futuristic technologies, drawing inspiration from historical myths and real-world archaeological discoveries. This trend continued into the late 20th and early 21st centuries, with an increasing focus on the ethical implications of resurrecting ancient technologies in futuristic settings. Authors embraced the complexities of integrating ancient artifacts and technologies within speculative worlds, often using them as catalysts for moral and existential dilemmas within their narratives. The historical overview of ancient tech in sci-fi literature thus reveals a rich tapestry of interconnected themes, spanning from the pioneers of the genre to contemporary storytellers, each contributing to the enduring allure of ancient technologies in the realm of science fiction.

Key Authors and Influential Works

The exploration of ancient technologies in science fiction literature has been greatly enriched by the contributions of numerous key authors and their influential works. These visionaries have not only incorporated elements of ancient technology into their narratives but have also reshaped our understanding of history and human ingenuity through their imaginative storytelling.

One of the most prominent figures in this realm is Isaac Asimov, whose prolific writing career intersected with themes of robotics, artificial intelligence, and futuristic societies. Asimov's iconic 'Robot' series, which delved into the ethical and existential dilemmas arising from advanced technology, often drew inspiration from the conceptualization of ancient automata and the age-old quest for creating sentient beings. His work laid the groundwork for exploring the intersection of ancient and future technologies within the human experience.

Additionally, the renowned author Philip K. Dick captivated audiences with his thought-provoking narratives that frequently featured speculative elements derived from ancient artifacts and enigmatic inventions. Through novels such as 'Do Androids Dream of Electric Sheep?' and 'Ubik,' Dick weaved intricate parallels between ancient mystique and technological evolution, challenging readers to contemplate the implications of resurrecting ancient technologies in a futuristic context.

Building upon this legacy, contemporary authors like Neal Stephenson and China Miéville have further expanded the literary landscape by infusing their works with profound reflections on the enduring allure of ancient technologies. Stephenson's 'Anathem' skillfully integrates historical philosophical concepts and speculative technologies, while Miéville's 'Perdido Street Station' masterfully incorporates elements of ancient alchemy and mechanical marvels into the fabric of his fantastical world-building, underscoring the enduring influence of ancient technologies on modern imaginings.

Moreover, pivotal works such as H.G. Wells' 'The Time Machine' and Mary Shelley's 'Frankenstein' have stood as canonical pillars within the genre, sparking a tradition of interweaving ancient technologies with speculative fiction and reshaping the way humanity's technological past is reimagined in literary realms.

By examining the rich tapestry of these influential authors and their seminal works, one can discern a lineage of creative exploration that transcends time, illuminating the perennial fascination with ancient technologies in science fiction literature and shaping our perceptions of the interplay between history and innovation.

Cinematic Adaptations: From Text to Screen

The translation of ancient technological themes from written works of science fiction to the cinematic medium has been a captivating and transformative process. Filmmakers often face unique challenges when adapting such material, as they must balance the preservation of the source material's essence with the visual demands and storytelling conventions of the big screen. For instance, classic works like H.G. Wells' 'The Time Machine' and Jules Verne's '20,000 Leagues Under the Sea' have undergone multiple adaptations, each reflecting the technological capabilities and cinematic trends of their respective eras. The transition involves careful consideration of how to bring historical and futuristic technologies to life in a visually compelling and intellectually stimulating manner. Cinematic adaptations offer an opportunity to enhance the audience's understanding and appreciation of ancient technologies through immersive visuals and dynamic storytelling. Beyond the portrayal of specific inventions or concepts, filmmakers also grapple with how to convey the societal and ethical implications of these technologies, often using dramatic tension and visual spectacle to capture the imagination of viewers. Furthermore, the use of special effects can elevate the portrayal of ancient technologies to new heights, creating awe-inspiring depictions that resonate with modern audiences. When successfully executed, cinematic adaptations of ancient technological themes not only entertain but also prompt reflection on the enduring relevance of past innovations and their potential impact on the future. By examining the evolution of these adaptations over time, from early silent films to contemporary blockbusters, one can gain insight into how societal attitudes towards technology have evolved alongside advancements in filmmaking techniques. This intersection of ancient themes and cinematic artistry serves as a testament to the enduring allure of speculative fiction, fostering a deeper appreciation for the complexities of human ingenuity across generations.

Major Themes and Archetypes

In the realm of science fiction literature, the exploration of ancient technologies has given rise to numerous major themes and archetypes that continue to captivate readers and inspire authors. One prevalent theme is the juxtaposition of advanced ancient civilizations with our contemporary understanding of technology. This dichotomy serves as a springboard for examining cultural and societal evolution, often questioning our assumptions about progress and innovation. Another prominent archetype involves the rediscovery and reactivation of forgotten or hidden ancient technologies, offering narratives of adventure and intrigue as characters unearth powerful artifacts and decipher complex mechanisms. These stories often weave in elements of archaeological quests and scientific exploration, lending an air of mystery to the ancient technologies in question. Additionally, the ethical implications of utilizing ancient technologies are frequently explored within science fiction narratives. Questions of responsibility, power dynamics, and unintended consequences arise as characters grapple with the potential impact of resurrecting long-lost inventions. Furthermore, the notion of ancient prophecies and myths intertwined with advanced technologies forms a rich tapestry of storytelling. Ancient legends and folklore often intertwine with futuristic concepts, leading to narratives that blend the mysticism of the past with the possibilities of the future. Alongside these overarching themes, various archetypes emerge within the portrayal of ancient technologies in science fiction. From sentient ancient AI guardians to time-traveling devices, these archetypes contribute to the rich diversity of stories that explore the intersection of ancient wisdom and futuristic innovation. Through the depiction of these major themes and archetypes, science fiction offers a compelling lens through which to contemplate the enduring allure of ancient technologies and their impact on the human experience.

Influence of Historical Myths on Sci-Fi Narratives

The influence of historical myths on science fiction narratives is a compelling subject that underscores the deep-seated relationship between ancient legends and futuristic imaginings. Throughout the annals of sci-fi literature and film, authors and creators have drawn inspiration from the rich tapestry of myths and folklore that have been passed down through generations. These timeless stories, often steeped in allegory and symbolism, continue to captivate the human imagination and provide a wellspring of creativity for science fiction writers.

One prominent aspect of this influence is the use of mythic themes and motifs as foundational elements in constructing sci-fi universes. Whether it's the hero's journey, the battle between good and evil, or the concept of destiny and fate, these archetypal narratives find resonance within science fiction, offering a familiar framework for exploring otherworldly landscapes and speculative technologies. The enduring allure of mythical storytelling also infuses sci-fi with a sense of timelessness, connecting contemporary audiences to the collective unconscious of humanity's cultural heritage.

Moreover, historical myths often serve as allegorical vehicles for addressing contemporary societal issues via the lens of speculative futures. By extrapolating and remixing ancient legends, sci-fi narratives confront present-day concerns, such as power dynamics, environmental crises, and the ethical implications of technological advancements. Imbuing these age-old myths with futuristic settings allows writers to engage with complex moral and philosophical dilemmas, offering readers and viewers a fresh perspective on pressing global challenges.

Another captivating facet of incorporating historical myths into science fiction is the reinterpretation of mythical characters and events through a futuristic prism. Whether it's reimagining ancient gods as advanced extraterrestrial beings or transposing legendary quests onto distant planets, such narrative adaptations blur the lines between fantasy and science fiction, inviting audiences to reconsider their perceptions of ancient lore in light of modern scientific understanding and speculative wonder. This fusion of classical mythology and futuristic speculation adds depth and nuance to sci-fi storytelling, prompting contemplation on the nature of belief, the origins of myth, and the untold mysteries of the cosmos.

Technological Anachronisms and Their Purpose

In the realm of science fiction, technological anachronisms play a pivotal role in shaping narratives and adding layers of complexity to the portrayal of ancient technologies. These anachronisms are deliberate departures from historical accuracy, introducing advanced futuristic elements into ancient settings. The purpose behind such creative manipulation serves multiple functions, contributing to the thematic depth and speculative nature of the genre.

Firstly, introducing technological anachronisms allows modern audiences to engage with familiar concepts within unfamiliar historical contexts. By interweaving advanced technology with ancient civilizations, writers and creators can explore universal themes such as power, morality, and human ingenuity in thought-provoking ways. This approach enables readers and viewers to contemplate the impact of contemporary advancements on historical epochs and ponder the potential repercussions of applying future technologies to bygone eras.

Moreover, technological anachronisms serve as a tool for allegorical storytelling, enabling authors to comment on contemporary societal issues through the lens of antiquity. By juxtaposing anachronistic devices alongside ancient civilizations, science fiction narratives can address pertinent topics like environmental degradation, socio-political hierarchies, and ethical dilemmas arising from scientific progress. This juxtaposition prompts audiences to reflect on the parallels between past and present challenges, fostering a deeper understanding of the human condition and the implications of rapid technological evolution.

Furthermore, the inclusion of technological anachronisms in science fiction allows for the exploration of hypothetical scenarios and alternative historical trajectories. By postulating advancements that diverge from actual historical developments, creators can construct speculative worlds that challenge conventional perceptions of ancient societies. This speculative element encourages critical thinking about the factors influencing technological progress and societal evolution, prompting audiences to reevaluate their assumptions about the trajectory of human civilization.

Additionally, the deployment of technological anachronisms cultivates a sense of wonder and awe, captivating audiences by presenting an intriguing blend of past and future. By seamlessly integrating futuristic innovations into ancient milieus, writers and filmmakers evoke a sense of mystery and fascination, instilling a sense of marvel at the potential capabilities of ancient civilizations. This narrative technique not only entertains but also inspires curiosity about the intersection of history and technology, stimulating interest in both fields among a diverse audience.

Ultimately, the purpose of incorporating technological anachronisms in science fiction is to enrich the imaginative landscape of ancient settings, enlivening historical narratives with speculative elements while encouraging contemplation about the enduring impact of technology on humanity. Through these creative manipulations, science fiction continues to expand the boundaries of storytelling, inviting audiences to embark on thought-provoking explorations of the past, present, and future.

Contemporary Science Fiction's Depiction of Ancient Inventions

In contemporary science fiction literature, the depiction of ancient inventions serves as a compelling lens through which authors explore the intersection of historical technology and futuristic imagination. It is within this context that writers often intertwine elements of ancient civilizations with technological advancements that are far beyond their time, crafting narratives that both captivate and provoke thought.

Whether it's incorporating advanced machinery in ancient mythological settings or envisaging sophisticated contraptions inspired by historical artifacts, contemporary science fiction writers have adeptly woven the allure of ancient inventions into their speculative worlds.

One prevalent theme delves into the concept of lost or forgotten technologies from ancient civilizations, reintroduced and repurposed in a futuristic human society. Authors often ponder over the ramifications of rediscovering such innovations, examining how they may disrupt existing power structures, socio-economic systems, or even redefine the understanding of past civilizations. By intertwining ancient inventions with speculative futures, these narratives prompt contemplation about humanity's progress and the potential implications of harnessing forgotten knowledge.

Moreover, contemporary science fiction frequently explores the ethical conundrums surrounding the utilization of ancient technologies. The portrayal of characters grappling with moral dilemmas, such as whether to resurrect ancient weaponry or deploy ancient healing methods, provides an avenue for reflection on the responsible application of powerful knowledge. This thematic exploration not only enriches the narrative depth but also encourages readers to critically evaluate the parallel ethical quandaries faced in our modern era, thus making the stories not only entertaining but also intellectually stimulating.

Furthermore, the depiction of ancient inventions in science fiction often intersects with archaeological and historical scientific research. Writers immerse themselves in the study of ancient cultures and technological achievements, amalgamating factual details with speculative extrapolations to construct intricate and plausible worlds. By aligning scientific principles with imaginative extrapolations, authors imbue their narratives with authenticity, enabling readers to vicariously experience the intersection of ancient ingenuity and futuristic progress.

Contemporary science fiction continues to present diverse and innovative portrayals of ancient inventions, demonstrating the enduring fascination with the confluence of history and speculative technology. Through these narratives, readers are afforded the opportunity to engage with complex ethical debates, confront the mysteries of historical innovation, and contemplate the potential impact of integrating ancient wisdom into the fabric of future societies.

Ethical Considerations and Theoretical Dilemmas

Exploring the portrayal of ancient technologies in science fiction inevitably leads to a discussion of the ethical considerations and theoretical dilemmas that arise from such depictions. As writers and readers delve into the speculative realm of ancient inventions, they are confronted with pivotal questions about the implications and consequences of integrating these imagined technologies into their narratives. One of the primary ethical considerations revolves around the potential romanticization or glorification of historical cultures through the lens of science fiction. Authors must be mindful of avoiding cultural appropriation or misrepresentation while portraying ancient technologies, respecting the integrity and authenticity of the cultures from which these inventions originated. Another critical ethical dilemma that surfaces is the responsible use of futuristic interpretations of ancient technologies. It becomes imperative for authors to ponder the potential impact of their fictional creations on public understanding and perceptions of history. Additionally, the proliferation of theoretical quandaries stemming from the intersection of advanced futuristic concepts with ancient technologies brings forth a host of moral and philosophical debates. These quandaries often revolve around themes of power dynamics, societal restructuring, and potential misuse of knowledge. Delving into hypothetical scenarios where ancient technologies are resurrected or reimagined in futuristic settings opens up complex ethical inquiries regarding the balance of progress and preservation, the distribution of newfound power, and the possible repercussions on global dynamics. Furthermore, the theoretical dilemmas emanating from the fusion of ancient technologies with modern advancements prompt contemplation on the ethical boundaries of technological innovation and historical extrapolation. Writers and readers are compelled to explore the potential ramifications and transformative effects of rekindling dormant ancient inventions within speculative contexts. As such, delving into these ethical considerations and theoretical dilemmas enhances the depth and complexity of science fiction narratives, encouraging critical reflection and robust engagement with the interplay between ancient technologies and ethical responsibility.

Future Trends in the Portrayal of Ancient Technologies

As we venture further into the digital age and witness rapid advancements in artificial intelligence, virtual reality, and biotechnology, the portrayal of ancient technologies in science fiction is poised to evolve significantly. One of the key future trends lies in the fusion of ancient wisdom with futuristic innovation. Authors and creators are increasingly exploring the idea of integrating ancient technologies, such as alchemy, mysticism, and lost civilizations, with cutting-edge futuristic concepts. This juxtaposition offers a rich tapestry for storytelling, providing a bridge between the past and the future that captivates audiences and challenges their perceptions. Additionally, the burgeoning field of archaeo-futurism is likely to shape the portrayal of ancient technologies in sci-fi literature and media. This interdisciplinary approach seeks to envision potential futures based on alternative interpretations of history and archaeology, creating speculative narratives that push the boundaries of imagination. Moreover, future trends indicate a heightened focus on the societal impact of resurrecting ancient technologies in a modern context. With increasing awareness of environmental sustainability and ethical considerations, storytellers are likely to explore the ramifications of rediscovering and implementing ancient knowledge in a world facing complex global challenges. The incorporation of non-western ancient technologies is also anticipated to be a significant trend in the portrayal of ancient technologies in science fiction. As diverse voices gain prominence in the genre, there is a growing emphasis on drawing from lesser-known historical contexts and indigenous perspectives, shedding light on advanced ancient technologies that have been overlooked or marginalized in traditional Eurocentric narratives. This inclusive approach not only enriches the tapestry of stories but also fosters a deeper appreciation for the ingenuity and innovation of diverse cultures throughout history. Furthermore, advancements in augmented reality and immersive storytelling technologies are expected to revolutionize the portrayal of ancient technologies in science fiction. The integration of interactive experiences that allow audiences to engage with reconstructed ancient inventions and machineries offers a new dimension to the exploration of antiquity in speculative fiction. By blurring the lines between the real and the imagined, these technologies hold the potential to transport audiences into immersive worlds where the wonders of ancient technologies come to life, creating an unparalleled sense of awe and fascination. In conclusion, the future portrayal of ancient technologies in science fiction is destined to soar to new heights, driven by a convergence of historical knowledge, technological innovation, and a growing appetite for captivating narratives that transcend time and space.

22

Resurrecting Ancient Technologies

Reviving the Past

Reviving ancient technologies is an endeavor that transcends mere historical curiosity. It is a quest to uncover the depth of human ingenuity and innovation from bygone eras, providing valuable insights into the technological prowess of ancient civilizations. The motivation for resurrecting these lost skills stems from a profound curiosity about the capabilities of our predecessors and their ability to harness the forces of nature to create remarkable artifacts and innovations. By unearthing and understanding these lost skills, we gain a deeper appreciation for the resourcefulness and creativity of ancient societies, fostering a renewed sense of respect for our shared human heritage. Furthermore, the revival of ancient technologies offers the potential for profound discoveries that can inform and inspire contemporary technological advancements. As we delve into the methods and materials utilized by ancient artisans and craftsmen, we unearth a treasure trove of knowledge that has the potential to revolutionize modern approaches to manufacturing, engineering, and design. Through this exploration, we are afforded a unique opportunity to bridge the gap between past and present, enabling us to learn from the wisdom of our predecessors and leverage that knowledge to shape the future. Additionally, the resurgence of ancient technologies holds promise in addressing modern challenges while promoting sustainability and environmental consciousness. By studying the practices of ancient cultures, which exemplified harmonious coexistence with the natural world, we stand to gain valuable insights that can guide our efforts in developing eco-friendly and resource-efficient technologies. In essence, reviving ancient technologies is not merely an academic or intellectual pursuit; it is a profoundly impactful venture that has the power to redefine our relationship with technology, history, and the world at large.

Technological Archaeology: Unearthing Lost Skills

Archaeology, as a discipline, has traditionally focused on uncovering physical remnants of past civilizations, such as artifacts, structures, and cultural remains. However, the emerging field of technological archaeology delves deeper into the investigation and resurrection of lost skills, techniques, and knowledge systems embodied in ancient technologies. This approach goes beyond the mere recovery of objects; it seeks to understand and reconstruct the methods, materials, and cognitive processes that underpinned the development and application of technology in antiquity. By unearthing these lost skills, contemporary researchers are afforded the opportunity to gain invaluable insights into the ingenuity and innovative prowess of our ancestors.

Technological archaeologists employ a range of interdisciplinary methods to unearth and interpret ancient skills. Utilizing advanced imaging technologies, such as 3D scanning and digital modeling, researchers can meticulously analyze and reconstruct ancient artifacts and mechanisms to discern their working principles and manufacturing techniques. Furthermore, material science and experimental archaeology play pivotal roles in this endeavor, enabling scholars to replicate historical processes and craft artifacts using authentic ancient methods and raw materials. These endeavors provide a hands-on understanding of the technical skills and challenges faced by ancient craftsmen and engineers, shedding light on their problem-solving capabilities and creative adaptability.

The study of ancient technologies extends beyond the physical reconstruction of artifacts; it involves delving into the social, economic, and cultural contexts in which these innovations flourished. Exploring the sociotechnical dimensions of antiquity allows for a nuanced comprehension of how technological practices were integrated within broader societal frameworks. For instance, the analysis of ancient manufacturing centers and trade networks provides crucial insights into the diffusion of technical knowledge across diverse regions and cultures. By unraveling these intricate webs of influence, technological archaeologists can map the interconnected nature of innovation in the ancient world, revealing cross-cultural exchanges and collaborative developments that shaped technological landscapes.

Moreover, the study of ancient technologies underscores the significance of indigenous knowledge systems and marginalized innovations, offering a holistic perspective that transcends Eurocentric narratives of technological progress. This inclusive approach recognizes the diversity of human ingenuity and dismantles biases that have historically overshadowed non-Western contributions to the global technological heritage. As a result, the pursuit of technological archaeology not only revitalizes forgotten skills but also fosters a more equitable and pluralistic understanding of the past, ensuring that the rich tapestry of human technological achievement is acknowledged and celebrated.

Case Study: The Antikythera Mechanism Reconstruction

The Antikythera Mechanism, a marvel of ancient engineering, has long captivated scholars and enthusiasts alike. Discovered in an ancient shipwreck off the coast of the Greek island of Antikythera in 1901, this intricate device dates back to the 1st century BC. Often referred to as the world's oldest computer, the Mechanism featured a complex system of gears and dials designed to track celestial movements, predict eclipses, and align with the cycles of the sun and moon. Given its level of sophistication, it stands as a testament to the advanced engineering skills of the ancient Greeks. However, for centuries, the true purpose and complexity of the Antikythera Mechanism remained a subject of debate.

In recent years, dedicated researchers and scientists have embarked on a remarkable journey to reconstruct and understand this ancient artifact. Through meticulous study of the surviving fragments and the application of advanced imaging technologies such as X-ray tomography and 3D scanning, they have unveiled the inner workings of the Mechanism, shedding light on its intricate design and functionality. This modern-day archaeological endeavor has revealed the ingenuity and technical prowess of the ancient craftsmen who developed this astonishing piece of machinery over two millennia ago.

The reconstruction of the Antikythera Mechanism has not only provided valuable insights into ancient Greek technology but has also sparked renewed interest in reviving and adapting ancient techniques for contemporary use. By gaining a deeper understanding of the craftsmanship and scientific knowledge embedded in this ancient device, researchers have opened doors to innovative applications in modern fields such as astronomy, mechanical engineering, and even computational science. The Antikythera Mechanism stands as a compelling example of the timeless relevance of ancient technologies and the potential for interdisciplinary collaboration between historians, archaeologists, and technologists.

As we delve further into the intricacies of this reconstruction, we uncover how the Antikythera Mechanism serves as an emblem of the enduring quest to decipher and harness the legacy of ancient technological achievements. Its significance extends beyond its historical context, offering a bridge between the past and present, inspiring a new wave of exploration and innovation. By unraveling the mysteries concealed within its gears and gearing systems, we gain not just a glimpse into the ancient world, but also a realization of the boundless possibilities that arise from resurrecting and comprehending ancient technologies.

Materials and Methods: Replicating Ancient Techniques

Replicating ancient techniques involves a multidisciplinary approach that merges historical research, archaeology, material science, and craft expertise. The challenge lies not only in understanding the technical aspects of ancient technologies but also in determining the available resources and tools at the time. By delving into the material composition, design, and manufacturing processes of ancient artifacts and structures, modern researchers aim to recreate and comprehend the ingenious methods utilized by our ancestors.

Archaeological excavations provide crucial insights into the materials used in ancient technologies. The composition and properties of metals, ceramics, and other remnants can offer vital clues about the craftsmanship of the era. Furthermore, interdisciplinary collaborations with historians, archaeologists, and material scientists allow for a deeper investigation into the technological advancements of the past.

In addition to understanding the materials, gaining proficiency in ancient methods is essential for their replication. This often involves learning traditional artisanal skills and experimental archaeology. By crafting objects using ancient tools and techniques, experts can gain firsthand experience and a practical understanding of the challenges faced by ancient engineers and artisans. Furthermore, through trial and error, modern craftspeople can uncover the intricacies of ancient techniques, shedding light on the level of skill, precision, and creativity required for their execution.

The role of technologies such as scanning electron microscopy, X-ray imaging, and 3D printing cannot be underestimated in this pursuit. These cutting-edge methods enable the detailed analysis and replication of intricate ancient artifacts while respecting the integrity of the original objects. Furthermore, the use of digital models and simulations allows researchers and craftsmen to visualize and replicate ancient technologies with enhanced accuracy and efficiency.

As we delve deeper into replicating ancient techniques, it becomes evident that ethical considerations must guide our endeavors. Preserving and respecting the cultural and historical significance of the practices and artifacts is paramount. Ethical guidelines should ensure that these processes do not exploit or harm indigenous knowledge and heritage. In parallel, collaboration with local communities and experts can provide invaluable insights into the social, cultural, and spiritual aspects embedded within ancient technologies.

Ultimately, the meticulous replication of ancient techniques not only serves to satisfy intellectual curiosity but also enriches our understanding of the evolution of human innovation. By resurrecting these methods, we honor the legacy of ancient civilizations, garner profound appreciation for their ingenuity, and pave the way for preserving this wealth of knowledge for future generations.

Digital and Virtual Reconstruction Technologies

In the quest to revive ancient technologies, the integration of digital and virtual reconstruction technologies has emerged as a vital avenue for exploring the capabilities of past innovations. By harnessing the power of advanced computer modeling, researchers can virtually recreate ancient artifacts and mechanisms, offering valuable insights into their functionality and design. Digital reconstructions allow for the meticulous examination of intricate details that may have been lost or degraded over time, providing a newfound understanding of the engineering prowess of our ancestors.

One of the key advantages of digital reconstruction is the capacity to test and refine hypothetical theories about the operation of ancient technologies. Through simulations and computational analysis, experts can explore various scenarios, leading to breakthroughs in comprehending the intended purpose and mechanics of enigmatic devices. This approach not only sheds light on historical mysteries but also inspires contemporary innovation by demonstrating the ingenuity of ancient engineers.

Moreover, the use of virtual reality (VR) and augmented reality (AR) technologies has the potential to bring ancient technologies to life in immersive experiences. By creating interactive simulations, individuals can engage with meticulously reconstructed ancient devices, gaining a deeper appreciation for the craftsmanship and sophistication of these creations. This digital resurrection transcends traditional academic study, fostering public interest and understanding of ancient technologies in ways that were previously unattainable.

The application of 3D scanning and printing technologies further enhances the scope of digital reconstructions, allowing for the physical manifestation of replicated ancient artifacts. These replicas serve as tangible educational tools, enabling researchers and enthusiasts to interact with accurate representations of historical objects. Additionally, by making these replicas accessible to wider audiences, museums and educational institutions can offer a more hands-on approach to learning about ancient technologies, encouraging curiosity and exploration.

However, it is essential to acknowledge the limitations and challenges inherent in digital and virtual reconstruction. While these technologies provide valuable insights, they are inherently reliant on the accuracy of available data and assumptions. As such, there is a need for rigorous research methodologies and continual refinement to ensure the fidelity of digital reconstructions. Furthermore, ethical considerations surrounding the appropriate use and interpretation of these reconstructions must be carefully addressed to avoid misrepresentations or misconceptions.

In conclusion, the integration of digital and virtual reconstruction technologies presents an invaluable opportunity to breathe new life into ancient technologies. By leveraging these innovative tools, we can unravel the mysteries of antiquity, inspire contemporary creativity, and preserve the legacy of our technological heritage for generations to come.

Ethical Considerations in Technology Resurrection

Resurrecting ancient technologies raises a complex array of ethical considerations that must be carefully navigated. As we delve into the realm of reviving technology from antiquity, we are faced with multifaceted questions regarding ownership, cultural heritage, and intellectual property rights. One of the foremost ethical quandaries pertains to the appropriation of ancient knowledge and the potential exploitation of indigenous wisdom for contemporary gain. By resurrecting ancient technologies, there is a risk of commercializing sacred traditions and profiting from heritage that belongs to specific communities or cultures. This demands a conscientious approach to ensure that the revival of ancient technologies respects the rights and interests of those whose ancestors devised these innovations. Furthermore, the act of resurrecting ancient technologies can inadvertently strip local communities of agency and autonomy if not executed with sensitivity and inclusivity. It is imperative to engage in dialogue with relevant stakeholders, including cultural custodians, scholars, and descendants of the original creators, to ensure that the resurrection process is conducted ethically and in mutual collaboration. Additionally, ethical considerations encompass the potential misrepresentation of historical truths and the distortion of cultural narratives. The revival of ancient technologies must refrain from perpetuating inaccuracies or propagating romanticized ideals, as this can lead to the dissemination of misinformation or the creation of misleading historical narratives. In tandem with this, we must recognize the importance of preserving the integrity of ancient technologies in their original societal contexts. Their resurrection should not overshadow or diminish existing cultural practices, beliefs, or customs but rather act as complementary additions to our understanding of the past. Moreover, ethical considerations extend to the responsible stewardship of revived technologies. This involves pondering the potential implications and unintended consequences of reintroducing ancient inventions into modern society. Central to this contemplation is an examination of the impacts on contemporary industries, economies, and environmental sustainability. The ethical use of resurrected technologies also necessitates a thoughtful analysis of its interoperability with current scientific advancements, ensuring that it contributes positively to progress without engendering adverse effects. As we confront these ethical dimensions, it becomes apparent that the resurrection of ancient technologies unveils a tapestry of intricate moral, cultural, and social inquiries. Addressing these complexities requires a concerted commitment to integrity, empathy, and collaborative engagement at every stage of the revival process.

Connecting Dots: Link from Myth to Machinery

Throughout history, numerous ancient myths and legends have showcased remarkable feats of technology that were far ahead of their time. From the mechanical marvels of Daedalus in Greek mythology to the advanced flying machines described in Hindu epics like the Ramayana, these narratives offer tantalizing hints at the existence of ancient technologies beyond our current understanding. The tantalizing question arises: could these myths be more than mere flights of fancy? Perhaps they are distorted echoes of real technological achievements that have since been lost to the ages. This section aims to explore the intriguing correlation between ancient myths and tangible machinery, sparking a profound reassessment of the boundaries between legend and reality. By delving into both classical and non-western traditions, we will unravel the intricate tapestry of ancient narratives, seeking to identify possible connections to actual technological innovations. Through careful analysis, we will scrutinize the descriptions and capabilities of mythological constructs, comparing them with known historical advancements to discern potential intersections between myth and machinery. Moreover, by consulting archaeological evidence and conducting interdisciplinary research, we endeavor to shed light on how these ancient allegories might be rooted in genuine technological achievements, long obscured by the passage of time and cultural transformation. Additionally, we will examine how contemporary scholars and innovators are drawing inspiration from these mythical accounts, recognizing the possibility of harnessing ancient wisdom to inform modern technological endeavors. This examination serves to elucidate the enduring allure of ancient lore and its potential relevance to present-day innovation. Ultimately, by connecting the dots between myth and machinery, this exploration seeks to expand our understanding of the deep interconnections between ancient narratives and real-world technological achievements.

Impact on Modern Engineering and Design

The resurrection of ancient technologies has exerted a profound impact on modern engineering and design, creating a dialogue between historical innovation and contemporary practice. By studying and reviving ancient technologies, modern engineers and designers have gained insight into the fundamental principles that underpin these innovations, leading to novel approaches in problem-solving and invention. Ancient technologies often demonstrate remarkable complexity and ingenuity, challenging the perception of historical societies as technologically primitive or unsophisticated. This realization has inspired a shift in mindset within engineering and design communities, fostering a deeper appreciation for the intellectual achievements of our ancestors. The integration of ancient techniques and knowledge into modern practices has resulted in the development of more sustainable, efficient, and resilient solutions. For instance, the application of ancient construction methods, such as those used in the creation of durable Roman concrete, has informed the advancement of contemporary building materials, contributing to increased durability and reduced environmental impact. Moreover, the study of ancient mechanisms, such as the Antikythera Mechanism, has fueled innovations in precision engineering and mechanical design, inspiring the development of intricate devices with practical applications in various fields. The symbiosis between past and present technologies has catalyzed interdisciplinary collaborations, allowing experts from different domains to leverage ancient wisdom in addressing complex challenges. Additionally, by embracing the concept of technological revival, designers have been able to reimagine products and structures through a historical lens, integrating cultural heritage and authenticity into modern designs. Such endeavors not only enrich contemporary aesthetics but also facilitate the preservation of diverse cultural narratives. The impact of resurrected ancient technologies extends beyond tangible artifacts and processes, influencing the philosophy and ethos of modern engineering and design. By acknowledging the enduring relevance of ancient innovations, practitioners are encouraged to adopt a holistic approach that considers the long-term consequences of their creations, fostering a more conscientious and responsible approach to technological advancement. Ultimately, the revitalization of ancient technologies has reinvigorated the landscape of modern engineering and design, promoting a dynamic synthesis of tradition and innovation that transcends temporal boundaries.

Preservation Through Revival: Securing Ancient Knowledge

As we delve into the realms of resurrecting ancient technologies, we inevitably encounter the significant aspect of preserving and securing ancient knowledge for posterity. The revival of these age-old technologies presents an unprecedented opportunity to safeguard and propagate the wisdom of our ancestors. Through the meticulous reconstruction and restoration of ancient devices and techniques, we not only gain insight into the ingenuity of bygone civilizations but also ensure that their invaluable legacy is preserved for future generations. By reviving and studying these ancient technologies, we can unlock a treasure trove of knowledge encompassing engineering prowess, scientific advancements, and cultural heritage. In doing so, we bridge the gap between antiquity and modernity, establishing a continuum of learning and innovation. The preservation of ancient knowledge through technology resurrection also contributes to the enrichment of academic disciplines.

It provides a platform for interdisciplinary collaboration, drawing expertise from fields such as archaeology, history, engineering, and material science. By amalgamating diverse areas of study, we broaden our understanding of ancient societies and cultivate a holistic appreciation for their contributions to human progress. Moreover, the revival of ancient technologies serves as a means of revitalizing cultural heritage. It fosters a deeper connection to the past and cultivates a sense of pride in ancestral achievements. This resurgence of cultural identity offers an opportunity for contemporary societies to reinvigorate their traditions and celebrate the enduring legacy of their predecessors. Furthermore, preserving ancient knowledge through the resurrection of technologies prompts ethical reflections on responsible innovation. As custodians of this inherited wisdom, we are entrusted with the ethical stewardship of ancient insights. This entails a conscientious approach to technological speculation and application, mindful of the potential impact on cultural narratives and historical authenticity. It underscores the responsibility to engage with this knowledge respectfully and responsibly, acknowledging its intrinsic value beyond mere scientific curiosity. The act of safeguarding ancient knowledge perpetuates a narrative of resilience and innovation, showcasing the adaptive spirit of humanity across epochs. It reinforces the notion that understanding and cherishing the legacies of the past are integral to shaping a meaningful and purposeful future. Thus, by securing ancient knowledge through the resurrection of technologies, we affirm our commitment to honoring the enduring wisdom of ancient societies while galvanizing the continued pursuit of knowledge and progress.

Conclusion: The Future of Resurrected Technologies

As we conclude our exploration of resurrecting ancient technologies, it becomes increasingly apparent that this endeavor holds great promise for the future. The revival of ancient knowledge not only offers a deeper understanding of our past but also provides invaluable insights for the advancement of modern technology and society as a whole. Looking ahead, the future of resurrected technologies presents multifaceted implications and opportunities that merit careful consideration.

One of the most prominent aspects to consider is the potential impact on innovation and technological development. By resurrecting ancient technologies, we gain access to forgotten techniques and approaches that could inspire novel solutions to contemporary challenges. The fusion of ancient wisdom with modern ingenuity has the potential to foster groundbreaking innovations across various fields, from engineering and architecture to medicine and sustainability.

Moreover, the resurgence of ancient technologies offers a unique opportunity to bridge the gap between traditional craftsmanship and cutting-edge advancements. As we revive and study age-old practices, we can identify sustainable and resource-efficient methods that align with current aspirations for environmental conservation and ethical production. This convergence of ancient and modern principles has the power to shape a more conscientious and sustainable technological landscape for future generations.

In addition to its technical implications, the revitalization of ancient technologies contributes to cultural preservation and historical appreciation. Through the reconstruction and utilization of ancient inventions, we pay homage to the ingenious minds of our ancestors and ensure the enduring legacy of their achievements. This preservation of cultural heritage enriches our collective narrative and fosters a deeper understanding of the diverse societies that have contributed to human progress.

However, the resurrection of ancient technologies also calls for thorough reflection on ethical considerations and responsible stewardship. While the pursuit of ancient knowledge brings forth immense possibilities, it necessitates careful ethical examination to avoid exploitation and misappropriation of indigenous practices or intellectual property. Sensitive engagement with communities and stakeholders is paramount to ensuring that the revival of ancient technologies respects cultural autonomy and values, while fostering mutually beneficial partnerships.

In essence, the future of resurrected technologies holds the promise of leading humanity towards a harmonious convergence of ancient wisdom and contemporary innovation. By embracing the knowledge of the past and integrating it into our quest for progress, we stand poised to cultivate a future where the legacy of our ancestors serves as a guiding beacon for transformative change and holistic advancement. As we embark upon this journey of rediscovery, let us tread mindfully and purposefully, recognizing the profound responsibility that accompanies the resurrection of ancient technologies.

23

Ancient Wisdom in Modern Technology

Bridging Time with Technology

Throughout the annals of human history, the sagacious wisdom of our predecessors has stood as a beacon for the burgeoning innovations of successive generations. As we navigate the complex labyrinth of modern technology, it becomes increasingly evident that the profound insights of ancient civilizations provide an indispensable framework upon which contemporary advancements are constructed. The unbroken thread of knowledge, woven meticulously through time, serves to bridge the yawning chasm between antiquity and the digital age. From the revolutionary mathematical principles of ancient Mesopotamia to the philosophical underpinnings of ethical design in ancient China, the continuum of ingenuity traverses epochs, steadfastly guiding our present endeavors with the timeless discernments of yesteryear. This relentless perpetuation of astuteness represents not merely a historical curiosity, but rather an enduring testament to the enduring relevance of ancestral techne. It is within this sacred continuum that we find ourselves, modern architects of progress, entrusted with the custodianship of traditions that have been nurtured throughout millennia. By acknowledging and embracing the unceasing dialogue between past and present, we honor the immutable legacy of wisdom left by our forebears, and in doing so, fortify the foundations upon which future technological marvels shall be erected.

Historical Insight and Contemporary Application

The study of ancient civilizations provides valuable insights that can be applied to modern technology. By examining the achievements and innovations of our ancestors, we gain a deep understanding of the foundational principles that underpin contemporary advancements. From the precision engineering of ancient temple complexes to the sophisticated water management systems of antiquity, there is much to be learned from historical technologies. Understanding how these ancient technologies were developed, deployed, and maintained can offer guidance for solving present-day challenges. Furthermore, exploring the societal context in which these technologies arose helps us appreciate the complex relationship between technology and culture. This historical insight allows us to identify enduring technological concepts that have stood the test of time, transcending generations and geographical boundaries.

Principles Borrowed from Antiquity for Modern Use

The principles borrowed from antiquity are a testament to the timelessness of human ingenuity. As we delve into the annals of history, we unearth pearls of wisdom that continue to resonate in our modern technological landscape. The ancient civilizations, with their astute observations and deep understanding of nature, have gifted us with invaluable insights. These principles, rooted in millennia-old knowledge, serve as guiding lights for contemporary innovation. Ancient philosophies, such as those found in Eastern traditions like Daoism and Confucianism, provide a holistic approach to harmonizing human activities with the environment, inspiring sustainable design and renewable energy practices. Furthermore, the architectural techniques employed by ancient cultures endure as the epitome of structural integrity and spatial efficiency, informing our present-day construction methodologies. The timeless principles of balance, proportion, and beauty espoused by ancient artisans influence not only art and design but also user experience and product aesthetics in modern technology. Utilizing materials available in abundance locally, as seen in the ancient civilizations, promotes regional sustainability and reduces environmental impact, offering a lesson crucial in today's globalized world. The practice of critical thinking and problem-solving strategies evident in ancient texts and treatises continues to inspire innovative approaches in various fields, including engineering, medicine, and mathematics. By examining the pioneering work of these early thinkers, we gain a profound understanding of the enduring relevance of ancient thought in contemporary contexts. These principles, honed over centuries, remind us that the most effective solutions often stem from the wisdom of the past, allowing us to amalgamate historical knowledge with cutting-edge advancements.

Case Studies: Ancient Solutions in Modern Problems

Throughout history, civilizations have grappled with complex challenges, devising ingenious solutions that stand the test of time. In the context of modern technology and innovation, the exploration of ancient solutions offers a treasure trove of knowledge, presenting viable answers to present-day problems. By delving into case studies that juxtapose ancient wisdom with contemporary issues, we can gain a profound comprehension of how our forebears addressed analogous dilemmas, thus inspiring innovative perspectives and solutions. One exemplary case study involves the sustainable agricultural practices of ancient civilizations such as the Indus Valley and Mayan cultures. These societies ingeniously implemented methods for maximizing crop yield while minimizing environmental impact. By studying their techniques, modern agronomists and environmental scientists can glean insights into developing sustainable farming systems that address the pressing concerns of food security and ecological sustainability. Furthermore, examining the ancient city planning and engineering accomplishments of societies like the Romans and the Mesopotamians provides valuable lessons for urban development today. The intricate water management systems of these ancient civilizations offer inspiration for creating resilient infrastructure capable of tackling contemporary issues related to water scarcity and urban flooding. Additionally, the incorporation of renewable energy strategies from historically advanced civilizations, such as the use of wind power by ancient Persians and Greeks, holds great potential for informing the renewable energy sector of today. By scrutinizing these historical examples, modern engineers and environmental researchers can devise innovative approaches to harness green energy sources for a more sustainable future. Equally compelling is the study of ancient medical practices and herbal remedies, which have inspired contemporary pharmaceutical research and holistic healthcare methodologies. Drawing upon the knowledge accumulated over centuries, modern medical practitioners can discern the efficacy of plants and traditional remedies, potentially offering alternative treatments for prevalent ailments. By exploring such case studies, we uncover the depth of ancient wisdom and its application to modern challenges, highlighting the invaluable reservoir of knowledge that awaits exploration and implementation.

Sustainable Practices Derived from Ancient Wisdom

Sustainability has become a prevailing concern in our modern world, with growing awareness of the finite nature of our planet's resources. Remarkably, many ancient civilizations embraced sustainable practices that resonate with contemporary environmental efforts. Drawing insights from these time-honored traditions provides invaluable wisdom for shaping our sustainable future.

Ancient cultures across the globe demonstrated a deep respect for nature, recognizing the interconnectedness of all living beings. For instance, the indigenous peoples of North America practiced sustainable agriculture through crop rotation and land stewardship, fostering biodiversity and preserving soil fertility. Similarly, the agricultural methods of ancient China, such as terrace farming and water conservation techniques, offer enduring lessons for efficient resource utilization.

The concept of sustainable design is not new; it can be traced back to the architectural marvels of ancient civilizations. These cultures ingeniously adapted their dwellings to harness natural elements, optimizing energy efficiency and thermal regulation. The passive cooling systems in ancient Middle Eastern architecture and the use of natural ventilation in South Asian structures exemplify the innovative integration of sustainable principles into building design.

Furthermore, ancient knowledge of herbal medicine and natural remedies aligns with the contemporary shift towards holistic healthcare and wellness. The traditional healing practices of indigenous societies, informed by the inherent healing properties of plants and natural elements, inspire modern approaches to botanical medicine and alternative therapies. Embracing this ancient wisdom underscores the potential for sustainable healthcare solutions rooted in nature.

The sustainable legacy of ancient wisdom extends beyond individual practices, influencing philosophical perspectives on human interaction with the environment. The ethical teachings of various ancient philosophies emphasize the importance of harmony and balance, instilling profound respect for ecological interconnectedness. This reverence for nature serves as a guiding principle for modern environmental ethics and sustainability initiatives.

In essence, the convergence of ancient wisdom with modern sustainable practices signifies a harmonious blend of tradition and innovation. By embracing sustainable practices derived from ancestral knowledge, we pave a path towards a regenerative future, where the wisdom of the past illuminates the trajectory of our shared humanity.

Enhancing Modern Medicine through Historical Learnings

The impact of ancient wisdom on modern medicine is a captivating testament to the enduring relevance of historical learnings in shaping the future. The integration of traditional practices and contemporary medical approaches has yielded fascinating insights and transformative outcomes in healthcare. Drawing from age-old knowledge systems, modern medicine has continually evolved to embrace holistic principles and alternative treatments that have long been part of ancient healing traditions.

Ancient cultures across the globe developed sophisticated medical practices that encompassed not only physical ailments but also mental and spiritual well-being. From Ayurveda in India to Traditional Chinese Medicine, these ancient systems recognized the interconnectedness of body, mind, and environment, an understanding that resonates deeply with the modern concept of integrative medicine. The emphasis on personalized care and preventive strategies found in ancient texts has profoundly influenced modern healthcare approaches, paving the way for a more patient-centric model.

Furthermore, the study of medicinal herbs and botanical remedies in ancient civilizations has provided a wealth of knowledge for modern pharmacology. Plants revered for their healing properties in ancient cultures have become crucial sources for developing pharmaceutical drugs that address a wide array of medical conditions. The synthesis of traditional herbal remedies with scientific advancements has led to breakthroughs in treating various ailments, illustrating the enduring efficacy of historical medicinal knowledge.

In addition to herbal medicine, ancient surgical practices have also inspired innovative techniques in modern surgery. The meticulous observations and procedures documented in ancient medical texts have served as a foundation for contemporary surgical advancements, contributing to safer and more effective surgical interventions. Learning from the precision and expertise of our ancient predecessors has elevated surgical practices to unprecedented levels of sophistication and success, ultimately benefitting patients worldwide.

Moreover, the exploration of ancient medical manuscripts has uncovered valuable insights into disease prevention, lifestyle modifications, and the cultivation of overall well-being. The profound understanding of anatomy and physiology elucidated by ancient scholars has informed contemporary medical education, enriching the curriculum with timeless wisdom and promoting a comprehensive understanding of human health and wellness.

As we continue to delve into the annals of history to glean medical insights from our ancestors, it becomes increasingly evident that the convergence of ancient wisdom and modern medicine holds immense promise for enhancing patient care, advancing treatment modalities, and fostering a harmonious coalescence of tradition and innovation.

Ancient Architectural Strategies in Modern Construction

The architectural prowess of ancient civilizations continues to inspire and inform modern construction practices. While the technological capabilities of our ancestors may seem limited compared to contemporary tools, their ingenious design strategies remain timeless and influential. From the precise masonry of the ancient Egyptians to the innovative structural feats of the Romans, the lessons learned from studying these ancient architectural marvels have greatly impacted modern construction. Understanding and implementing these strategies not only pays homage to our history but also offers practical benefits in our present-day built environment. One prominent example lies in the concept of load-bearing structures, an ancient architectural technique that has seamlessly integrated into modern construction. Ancient engineers and builders mastered the art of distributing weight through strategic placement of materials, which is a fundamental principle evident in today's architectural designs. By harnessing this age-old wisdom, architects and engineers continue to create durable, efficient, and aesthetically pleasing structures. Moreover, the use of sustainable building materials and techniques, such as those employed by ancient civilizations, serves as a guiding light for contemporary efforts in eco-friendly construction. Incorporating natural elements like clay, wood, and stone–materials widely favored by our ancestors–showcases a commitment to both historical reverence and environmental responsibility. The longevity and resilience of many ancient structures, including temples, amphitheaters, and aqueducts, stand as testaments to the efficacy of their architectural approaches. These enduring monuments provide valuable insights into durability, longevity, and climate adaptability, essential considerations in today's construction industry. Furthermore, the philosophical underpinnings of ancient architecture, emphasizing harmony with nature and the surrounding environment, resonate deeply with the principles of sustainable and holistic design embraced in modern times. This alignment of ethos transcends millennia, demonstrating the enduring relevance of ancient architectural wisdom in shaping contemporary construction practices. By embracing these time-honored strategies, our modern built environment can strike a harmonious balance between innovation, functionality, and respect for heritage, resulting in structures that honor the past while propelling us towards a more sustainable future.

Time-tested Philosophies Guiding Technological Ethos

Throughout the ages, civilizations have developed intricate philosophies that embody their approach toward life and innovation. These time-tested philosophies have often been intertwined with technological pursuits, shaping the ethos of societies and driving progress in diverse domains. The integration of ancient wisdom into modern technological ethos is a compelling phenomenon that resonates across cultures and disciplines.

The venerable wisdom of antiquity continues to guide contemporary technological advancements, offering valuable insights for ethical decision-making, sustainability, and inclusive design principles. By delving into the philosophical underpinnings that informed ancient crafts and ingenuity, we can discern enduring principles that serve as ethical compasses in the fast-paced world of modern technology.

One such principle is the emphasis on harmony with nature. Many ancient philosophies upheld the belief that humans are intrinsically connected to the natural world, advocating for responsible stewardship of the environment. This reverence for nature's balance and resilience remains pertinent today, inspiring sustainable technologies and eco-conscious innovations aimed at mitigating environmental impact. The timeless wisdom of living in harmony with nature serves as a touchstone for modern technological ethos, fostering the development of green technologies and ethical frameworks that prioritize planetary well-being.

Furthermore, the concept of holistic well-being, deeply embedded in ancient philosophical traditions, has found resonance in the modern tech landscape. The contemplation of human flourishing and societal welfare as central tenets of progress has led to the incorporation of wellness-centric designs in contemporary technologies. From user experience to product development, this emphasis on holistic well-being fuels an ethical approach to innovation and underscores the importance of creating technologies that empower and uplift individuals and communities.

Additionally, ancient philosophical traditions espoused principles of inclusivity and equity, seeking to address the needs of diverse populations within societal frameworks. Today, these ideals find expression in the pursuit of accessible and inclusive technological solutions, where the ethos of universal design and equitable access informs the development of products and services. The infusion of age-old philosophical concepts of fairness and empathy into technological innovation propels the creation of tools and systems that cater to the diverse needs of global audiences, promoting inclusivity and social cohesion.

By integrating these time-tested philosophies into the fabric of modern technological ethos, we embark on a journey of ethical innovation that harmonizes the wisdom of the past with the imperatives of the future. This convergence enriches the technological landscape, infusing it with profound ethical considerations, sustainability imperatives, and a steadfast focus on collective well-being. As we navigate the complexities of the digital age, drawing inspiration from ancient philosophies empowers us to forge a more enlightened and conscientious path in our technological endeavors.

Education and Knowledge Transfer Across Millennia

The exchange of knowledge and ideas has been a defining characteristic of human civilization throughout history. From the ancient libraries of Alexandria to the modern digital repositories, the preservation and dissemination of knowledge have played a crucial role in societal progress. This section delves into the enduring legacy of education and the transfer of knowledge across millennia.

Ancient civilizations recognized the importance of education and established institutions to impart wisdom to future generations. Through the oral tradition, written texts, and apprenticeship systems, invaluable knowledge was passed down from elders to youths. The ethos of preserving and transmitting knowledge has permeated various cultures, fueling the flame of innovation and discovery.

One of the most remarkable aspects of knowledge transfer across millennia is the continuity of key principles and concepts. Diverse fields such as mathematics, astronomy, medicine, and philosophy have seen foundational ideas endure through time, shaping the development of modern disciplines. Lessons learned in managing resources, organizing societies, and harnessing natural forces continue to find relevance in contemporary contexts.

The advent of digital technology has revolutionized the accessibility and scope of education and knowledge transfer. Online platforms, virtual classrooms, and e-learning resources have expanded the reach of education beyond geographical boundaries. This interconnectedness enables the global exchange of ideas, facilitating cross-cultural learning and collaboration—a testament to the timeless pursuit of knowledge transcending physical borders.

Furthermore, the reinterpretation of ancient texts and artifacts through modern scholarly lenses provides insight into the complexities of ancient knowledge systems. Interdisciplinary studies and collaborative research efforts have unraveled enigmatic practices and techniques, shedding light on the sophistication of ancient educational paradigms. By bridging the gap between the past and present, contemporary scholars continue to uncover new layers of understanding and appreciation for the wisdom of our ancestors.

In conclusion, the sustained relevance of education and knowledge transfer across millennia underscores the enduring legacy of human intellectual pursuits. As we stand on the shoulders of our forebears, it is imperative to recognize the invaluable contributions of ancient wisdom to our current repository of knowledge. Embracing this interconnected tapestry of enlightenment fosters a richer understanding of our collective heritage and empowers us to chart a path towards a more enlightened future.

Conclusion: Integrating Ancestral Insights into Future Technologies

The journey through the annals of history has illuminated the invaluable knowledge and wisdom that ancient civilizations possessed. As we stand at the crossroads of present and future, it is imperative to recognize the enduring relevance of these age-old insights in shaping the trajectory of technological advancement. The convergence of ancestral wisdom with contemporary innovation presents a transformative opportunity, one that holds the potential to redefine our approach to progress and sustainability.

In integrating ancestral insights into future technologies, we are presented with a nuanced tapestry of concepts, principles, and practices that have withstood the test of time. It is not merely a retrospective exercise but a proactive endeavor to tap into the reservoir of accumulated human experience. By assimilating these intricate threads of knowledge, we pave the way for a more holistic and conscientious evolution of technology.

This integration necessitates a concerted effort to explore the underlying philosophies that underpin ancient wisdom and translate them into actionable frameworks for modern technological development. It demands a departure from the myopic view of progress solely based on novelty, instead calling for a discerning gaze towards time-honored approaches that promote harmony, efficiency, and sustainability.

Moreover, this amalgamation of past and present presents a unique opportunity to foster interdisciplinary collaboration, transcending the boundaries of specialization and embracing a synthesis of diverse perspectives. To effectively integrate ancestral insights into future technologies, we must cultivate an ecosystem where historians, archaeologists, technologists, engineers, and ethicists converge to enrich the discourse and infuse innovative solutions with historical depth.

The implications of this integration extend beyond the realms of technological innovation; they permeate societal structures, ethics, environmental stewardship, and global interconnectedness. Therein lies the transformative potential of weaving together ancient wisdom and modern technology—a prospect that offers a harmonized vision for progress rooted in the collective wisdom of humanity's journey through time.

As we embark on this endeavor, it is essential to acknowledge the responsibility that comes with harnessing ancient wisdom. With great reverence for the teachings of our predecessors, we must approach this integration with humility and inquisitiveness, cognizant of the need for ethical considerations, responsible stewardship, and equitable dissemination of advancements.

In conclusion, the seamless integration of ancestral insights into future technologies symbolizes a pivotal juncture in the human narrative. It symbolizes a departure from the linear trajectory of progress towards a cyclical, inclusive, and sustainable paradigm—one that honors the legacies of the past while stepping decisively into the promise of the future.

24

Ancient Technology in Contemporary Art

Introduction to Artistic Transformations

Throughout history, art has served as a visual narrative of humanity's evolution, reflecting the collective consciousness and innovative spirit of respective eras. By delving into the confluence of ancient technologies and contemporary artistic expressions, we embark on a mesmerizing journey that transcends time, weaving a seamless tapestry that connects antiquity with the present day. The breath-taking achievements and advancements of ancient civilizations underscore their unparalleled ingenuity, offering timeless inspiration for modern artists seeking to infuse their work with a sense of aesthetic continuity spanning centuries. By forging this bridge between antiquity and the contemporary world, artists have harnessed the essence of bygone technologies to breathe new life into their creations, thereby enabling audiences to experience the allure of ancient wonders with fresh perspectives. This transformative process not only revitalizes historical legacies but also imbues art with captivating layers of meaning, invoking a profound sense of interconnectedness across epochs. As we traverse these artistic transformations, intricate patterns emerge, revealing the enduring impact of ancient technologies on shaping creative landscapes. Through insightful exploration, we uncover how artisans draw from ancient methods, materials, and designs, contributing to a rich artistic tapestry that extends beyond temporal boundaries. Embracing and reimagining the skillful applications of ancient techniques empowers contemporary artists to infuse their works with a depth of historical resonance that captivates and resonates with audiences, fostering a deeper appreciation for the dialogue between past and present. In essence, this chapter endeavors to unravel the symbiotic relationship between ancient technology and contemporary artistic expressions, encapsulating the essence of human creativity and innovation, while perpetuating the legacy of our ancestors' remarkable achievements.

The Aesthetic of Antiquity: A Contemporary Context

In the realm of contemporary art, the profound influence of ancient aesthetic principles is palpable, resonating across diverse mediums and expressions. This enduring appeal lies in the innate human inclination to seek beauty and meaning, transcending temporal boundaries, while honoring the timeless allure of antiquity. The fusion of modern artistic sensibilities with elements derived from ancient cultures bears testament to the living legacy of historical creativity.

Ancient aesthetics encapsulate a remarkable blend of form, function, and philosophical underpinnings that continue to captivate artists and audiences alike. Whether conveyed through the delicate curves of Hellenistic sculptures or the intricate geometric patterns of Islamic art, these timeless visuals serve as wellsprings of inspiration for contemporary creators. Moreover, the reinterpretation of ancient symbols and archetypes imbues present-day artworks with a depth of historical connection, fostering a universal language that speaks across epochs.

For artists seeking introspection, the study of ancient aesthetics offers a treasure trove of wisdom, providing insights into the societal values, spiritual beliefs, and technological innovations of bygone eras.

This journey through time enables a contextual reframing of modern artistic endeavors, embedding them within the narrative continuum of human creativity. Furthermore, the interplay between tradition and innovation becomes manifest as artists infuse ancient motifs with avant-garde techniques, forging an evocative dialogue between the echoes of antiquity and the pulse of contemporary expression.

It is essential to acknowledge the challenges inherent in harmonizing ancient aesthetics with the dynamism of modern art. Sensitivity to cultural authenticity and respectful homage to historical contexts must underscore this fusion, ensuring that the synthesis remains a celebration of shared heritage rather than an appropriation of cultural identity. By navigating this delicate balance, artists can weave narratives that resonate with global audiences, enriching collective consciousness with the enduring aesthetic wisdom of antiquity.

Ultimately, the integration of ancient aesthetics with contemporary artistic expression serves as a testament to humanity's unbroken creative lineage, tendering a bridge across centuries and civilizations. With mindful homage to the masterpieces of antiquity, artists navigate the complexities of temporal dialogue, crafting narratives that invoke the eternal allure of historical beauty within the evolving tapestry of modern art.

Sculptural Symmetry: From Stone Carving to Digital Rendering

In the realm of artistic expression, the evolution of sculptural techniques from ancient civilizations to the contemporary digital age reflects a rich tapestry of human creativity and innovation. The age-old practice of shaping stone into intricate forms has seamlessly transitioned into the virtual realm, where artists navigate the digital landscape with precision and imagination. This metamorphosis, from hand-carved statues to virtual sculptures, exemplifies the enduring resonance of ancient technology in the modern world.

The craftsmanship of ancient sculptors, evident in revered masterpieces like the Great Sphinx of Giza and the Parthenon marbles, serves as an enduring testament to the artistry and technical prowess of bygone eras. These timeless works, meticulously crafted through laborious chiseling and carving, capture the essence of their respective cultures and beliefs, offering a window into the technological achievements of early civilizations. Such tangible manifestations of ancient sculptural prowess continue to inspire contemporary artists seeking to infuse their creations with a sense of historical continuity and cultural depth.

The advent of digital sculpting software and 3D modeling technologies has revolutionized the artistic process, effectively bridging the gap between tradition and innovation. Artists now wield virtual chisels and brushes, allowing them to sculpt and refine ethereal forms with unprecedented intricacy and detail. By leveraging these digital tools, creators can explore the boundless possibilities of sculpting without the constraints of physical materials, transcending the limitations imposed by stone or marble. This convergence of traditional sculptural principles with cutting-edge digital platforms has ushered in a new era of artistic expression, enabling the realization of intricate designs previously inconceivable through conventional methods.

Furthermore, the integration of digital sculpting techniques with advancements in additive manufacturing has propelled sculpture into the realm of multidimensional art forms. The seamless translation of digital sculptures into tangible objects via 3D printing technologies represents a testament to the enduring legacy of ancient craftsmanship in contemporary art. This fusion of ancient sensibilities with modern methodologies not only expands the horizons of artistic creation but also revitalizes the timeless allure of sculptural symmetry for present and future generations.

As contemporary artists continue to explore the profound heritage of sculptural traditions, the synthesis of ancient craftsmanship with digital ingenuity establishes a harmonious dialogue between past and present. By navigating the dynamic juxtaposition of tactile artistry and virtual precision, creators honor the lineage of sculptors while charting new frontiers of aesthetic exploration. Consequently, the evolution from stone carving to digital rendering epitomizes the enduring magnificence of sculptural symmetry, transcending temporal boundaries to unite the artistic endeavors of antiquity with the boundless possibilities of contemporary expression.

Ancient Motifs in Modern Painting

In the realm of contemporary art, the influence of ancient motifs on modern painting is a captivating expression of timelessness. Across a multitude of cultures and civilizations, ancient art has left an indelible mark on the aesthetic sensibilities of artists throughout history. The resurgence of interest in these ancient motifs in modern painting serves as a testament to their enduring appeal. Key to this phenomenon is the evocation of mythology, symbolism, and cultural narratives that intertwine the ancient and the modern. Artists draw inspiration from diverse sources such as Egyptian hieroglyphs, Mesopotamian reliefs, Greek pottery, and Mesoamerican murals, infusing their work with a rich tapestry of historical significance. By incorporating these timeless elements into their paintings, artists bridge the chronological chasm between antiquity and the present, creating a visual dialogue between epochs. Whether through vibrant hues or subdued palettes, the application of ancient motifs in modern painting conveys a profound respect for the craftsmanship and artistic vision of our predecessors. Moreover, this fusion of past and present serves as a contemplative exploration of the human experience, reinterpreting ancient narratives within the context of contemporary society. As the convergence of tradition and innovation unfolds on the canvas, viewers are invited to ponder the resonance of ancient themes in their own lives, fostering a deeper connection to the enduring legacy of antiquity. From allegorical depictions of ancient deities to abstract reinterpretations of ancient landscapes, modern painters exalt the timeless allure of antiquity while engaging with the complexities of the modern world. The interplay of traditional techniques and cutting-edge artistic methods further amplifies the significance of this artistic discourse, highlighting the perpetuity of human creativity and ingenuity. By weaving the threads of antiquity into the fabric of their paintings, artists ignite a dialogue that surpasses temporal boundaries, inviting audiences to reflect on the continuum of human expression and cultural heritage. Ultimately, the integration of ancient motifs in modern painting underscores the universality of human experience and the unyielding relevance of ancient art, perpetuating a timeless narrative that transcends the confines of temporal constraints.

Artistic Integration of Ancient Engineering Principles

Ancient engineering principles have left an indelible imprint on the world of contemporary art, offering a rich tapestry of inspiration for artists and creators. The seamless fusion of artistic expression with age-old techniques has given rise to a diverse range of artworks that pay homage to the mastery of ancient engineers while encapsulating the ethos of modernity. From awe-inspiring sculptures to avant-garde installations, artists are increasingly embracing the challenge of intertwining traditional engineering prowess with their creative vision to produce evocative and impactful pieces. One can witness this synthesis in various forms, where sculptures resonate with the harmonious proportionality championed by ancient architects, and installations echo the intricate mechanisms symbolized in ancient machinery. Moreover, the integration of ancient engineering principles into art not only serves as a tribute to the ingenuity of the past but also acts as a catalyst for contemplation on the ever-evolving relationship between human innovation and the enduring legacy of antiquity. Artists often harness the principles of balance, stability, and precision that defined ancient engineering to infuse their creations with a sense of timelessness and transcendence. This deliberate infusion of engineering principles underscores the enduring relevance of ancient wisdom in an ever-changing artistic landscape, promoting a dialogue between the innovations of yesteryears and the boundless possibilities of tomorrow. As such, the marriage between ancient engineering and contemporary artistic forms transcends mere aesthetic appeal; it embodies a profound testament to the enduring spirit of humanity's quest for beauty, meaning, and connection. The result is a remarkable convergence of tradition and modernity, illustrating how the reservoir of historical technology continues to inspire and inform the myriad expressions of creativity in the present day.

Rebirth of Classical Architectural Techniques in Modern Designs

The integration of classical architectural techniques from ancient civilizations into modern designs marks a pivotal resurgence of historical significance in contemporary architecture. Drawing inspiration from the structural marvels crafted by our forebears, architects and designers today are reimagining, reinterpreting, and integrating these timeless techniques into their works with remarkable creativity and innovation. The symmetrical precision and grandeur of ancient buildings, such as the Parthenon in Greece or the Pantheon in Rome, have become enduring touchstones for contemporary architects seeking to infuse their creations with a sense of timeless beauty and cultural depth. Through the study of ancient construction methods, materials, and principles, modern practitioners are able to pay homage to the classical past while simultaneously pushing the boundaries of architectural possibility. Embracing the craftsmanship and artistry of antiquity, architects are reviving lost techniques, such as stone masonry and intricate wood joinery, to create structures that marry historical elegance with contemporary functionality. Some architects experiment with replicating ancient building methods using modern technology like 3D printing, combining the best of both worlds to achieve stunning results. Additionally, the revival of traditional materials such as marble, limestone, and terracotta not only reflects a commitment to sustainable construction but also honors the legacy of ancient builders.

Furthermore, archaeological discoveries and detailed examinations of ancient ruins provide architects with invaluable insights into the complex engineering feats employed by our ancestors. By incorporating this knowledge into their projects, modern architects are able to evoke the awe-inspiring majesty of historical edifices while embracing the demands of present-day society. The resurgence of classical architectural techniques in modern designs serves as a testament to the enduring influence and relevance of ancient technology, propelling architectural innovation into a harmonious dialogue between past and present.

Technology and Textile: Weaving Historical Threads into Modern Fabric

The interplay between ancient technology and contemporary textile arts unveils a rich tapestry of innovation and creativity. Drawing inspiration from the meticulous craftsmanship of bygone civilizations, modern fabric design has embraced a seamless fusion of tradition and technology. Ancient weaving techniques, once confined to the annals of history, have been revitalized through the utilization of cutting-edge machinery and digital processing. This symbiotic relationship between the past and the present has laid the foundation for a captivating narrative that celebrates the enduring legacy of ancient textile technologies.

Embracing the ethos of preservation, textile artists pay homage to their predecessors by resurrecting time-honored methods in tandem with sophisticated modern tools. Intricate patterns and motifs derived from historical textiles are reborn in contemporary fabrics, echoing the skill and artistry of ancient weavers. The convergence of ancient and modern materials gives rise to an eclectically textured landscape, where the whispers of antiquity are woven into every fiber.

Furthermore, the integration of state-of-the-art looms and computer-aided design software has redefined the parameters of textile creation, allowing for intricate detailing and precision that would have mesmerized artisans of old. The marriage of ancient inspirations with modern technological capabilities paints a vivid picture of cultural continuity and artistic evolution. It is within this realm of innovation that ancient textile traditions find resonance in the bustling heartbeat of the present day, transcending temporal boundaries.

As contemporary creators delve into the treasure trove of ancient textile methodologies, they meld history with visionary ingenuity, birthing iconic pieces that pay homage to their illustrious ancestry while charting new frontiers of expression. By infusing ancient dyeing techniques and material manipulation concepts with contemporary aesthetics, they forge a bridge between antiquity and modernity, projecting a symbiosis of timeless allure and avant-garde allure. The resulting fabric masterpieces stand as living testaments to the enduring relevance and adaptability of ancient textile technologies in the ever-evolving world of contemporary art.

Ancient Ceramics and their Influence on Modern Ceramic Arts

The cultural legacy of ancient ceramics reverberates through the timelines, weaving a narrative of tradition and innovation that continues to inspire contemporary ceramic arts. Ancient civilizations such as the Greeks, Romans, Chinese, and Indigenous peoples across the globe mastered the art of pottery, elevating it into a form of expression and utility. Their exquisite craftsmanship and technical prowess in working with clay and glazes laid the foundation for modern ceramic arts. Through archaeological discoveries and scholarly studies, we have gained invaluable insights into the techniques and aesthetics of ancient ceramics, which have significantly impacted the direction of contemporary artistic endeavors. The use of kilns, wheel-throwing, and intricate decorative methods employed by our forebears serve as a reservoir of knowledge that enriches and informs the evolution of modern ceramic artistry. Understanding the historical development of ceramic materials and forms enables artists to pay homage to these ancestral practices while reviving them in a modern context. The timeless appeal of ancient pottery shapes, motifs, and surface treatments finds resonance in the creations of today's ceramicists, establishing a bridge between the past and present. Furthermore, the utilization of ancient firing techniques and naturalistic embellishments offers contemporary potters a deeper connection to the earth and its elements, fostering a sense of continuity with our ancestors. As custodians of cultural heritage, modern ceramic artists treasure the essence of ancient techniques, integrating them into their work to preserve and propagate the artistic traditions of past civilizations. By infusing their pieces with the spirit of ancient ceramics, contemporary artists honor the craft's lineage while infusing it with fresh interpretations and expressions. Through this amalgamation, they demonstrate reverence for the resourcefulness and creativity of our predecessors, ensuring that the profound legacy of ancient ceramics thrives in the diverse and vibrant tapestry of modern ceramic arts.

Preservation through Art: Ancient Technology as a Muse

Art, through various mediums, has been instrumental in preserving and celebrating ancient technologies. It serves as a captivating bridge that connects the past to the present and ensures that the wisdom of our ancestors endures through the ages. One profound way in which contemporary artists pay homage to ancient technology is by drawing inspiration from historical artifacts and innovative practices. By incorporating these influences into their work, artists not only honor the ingenuity of ancient civilizations but also breathe new life into traditional techniques. Through this creative process, they infuse ancient crafts with modern relevance, ensuring that these methods and their significance remain vibrant and meaningful in today's world. Furthermore, the use of ancient technology as a muse for art helps to raise awareness about the historical significance and contributions of early civilizations. It brings attention to the exceptional skills and knowledge possessed by ancient artisans and scientists, shedding light on their capabilities in fields such as metallurgy, ceramics, architecture, and more. This spotlight on ancient technology invigorates interest in preserving traditional craftsmanship and encourages continued exploration into the innovative solutions developed by our ancestors. Additionally, artistic interpretations of ancient technology provide viewers with a tangible connection to the past, invoking a sense of awe and curiosity that fosters an appreciation for the journey of human progress. It prompts reflection on the timeless relevance of ancient techniques and inspires contemplation of how these methodologies continue to influence contemporary society. Ultimately, by infusing ancient technology into their art, creators play a pivotal role in safeguarding the legacy of our predecessors, ensuring that their groundbreaking achievements are celebrated and remembered. This preservation through art not only enriches modern culture but also cultivates a deeper understanding and reverence for the marvels of ancient technology, setting the stage for future endeavors that build upon this foundation.

Summary and Future Prospects

The contemporary integration of ancient technology into artistic pursuits presents a confluence of historical homage and modern innovation. Through the conscientious preservation of ancient techniques, artists today are not merely reviving relics of the past, but rather perpetuating a narrative that transcends time. This summary delves into the overarching impact of this symbiotic relationship and contemplates the future prospects it offers.

In summarizing the preceding discussions, it is evident that artists have been harnessing the spirit of ancient technology as a muse, infusing their creations with the essence of bygone eras. By incorporating age-old methods and aesthetics into contemporary art, they pay homage to the ingenuity of early civilizations, honoring their legacy in a way that surpasses conventional historical documentation. The resurgence of these ancient techniques in modern artworks also serves as a catalyst for generating renewed interest and appreciation for antiquity, effectively bridging the chasm between past and present.

Looking towards the future, the prospects for this amalgamation are both exciting and promising. As technological advancements continue to burgeon, there exists an unprecedented opportunity to further blend the innovative prowess of ancient civilizations with the cutting-edge capabilities of the present day. This convergence may lead to the development of entirely novel artistic forms that push the boundaries of creativity and expression. Additionally, the perpetuation of ancient technology through contemporary art can serve as a conduit for cultural preservation, safeguarding the essence of our heritage amidst the ever-changing landscape of human endeavor.

Moreover, the collaborative synergy between artists, historians, and technologists holds the potential to catalyze deeper explorations into ancient methodologies, inspiring interdisciplinary dialogues that enrich both the artistic and scholarly domains. By fostering collaborative initiatives that bridge the chasm between different fields of expertise, we pave the way for a renaissance of ancient wisdom, wherein tradition and innovation harmoniously coalesce to forge previously uncharted frontiers.

In conclusion, the crossroads of ancient technology and contemporary art not only begets a celebration of human accomplishments through the ages but also paves a pathway towards a future where the amalgamation of old and new heralds a renaissance that transcends temporal confines. By embracing the heritage of our forebears and reshaping it within the crucible of modernity, artists stand as torchbearers of an eternal flame that illuminates the corridors of our collective legacy.

25

Ancient Technology in Space Exploration

Historical Context and Significance

Ancient civilizations across the globe demonstrated a profound understanding of cosmology and celestial mechanics. Their astronomical endeavors were not only pioneering but also crucial precursors to the modern exploration of space. The significance of ancient technological insights for contemporary space research cannot be overstated, as it forms the foundation upon which our current understanding of the cosmos rests. By delving into the historical context of these early developments, we gain valuable perspectives that enrich our approach to space exploration. The ancient wisdom and technological innovation offer a bridge between past and present, helping us comprehend the universe in ways that go beyond the limitations of our current knowledge.

The legacy of ancient cosmological concepts permeates through the annals of history, captivating the imagination of future generations. Civilizations such as the Babylonians, Egyptians, Greeks, Indians, and Chinese, among others, made remarkable strides in comprehending the mechanics of celestial bodies and cosmic phenomena. Their astrological pursuits, precise astronomical observations, and innovative instruments bespeak a pursuit of knowledge that transcended the confines of their time. These early pioneers laid a resilient groundwork for the scientific principles that propel our contemporary space missions.

By acknowledging the profound significance of ancient cosmological insights, we honor the intellectual endeavors of our forebears and recognize the enduring relevance of their contributions. Moreover, by studying the navigational aids, observational tools, and cosmological theories of antiquity, we gain a holistic perspective that fuels our advancements in space science. Integrating these historical precedents into modern research frameworks enables us to tap into the wealth of knowledge accumulated over centuries and expand our understanding of the cosmos, transforming ancient visions into contemporary realities.

Fundamentally, the historical context of ancient cosmology provides a testament to the timeless curiosity and ingenuity of humankind. It offers a compelling narrative that underscores the depth of human intellect and perseverance, inspiring today's space researchers to push the boundaries of our knowledge further. Recognizing the historical significance of ancient technological insights for modern space research not only pays homage to our predecessors but also propels us forward in our quest to unlock the mysteries of the universe.

Concepts of Cosmology in Ancient Civilizations

Ancient civilizations, often revered for their architectural wonders and cultural achievements, also harbored sophisticated cosmological concepts that continue to captivate modern scholars. The pursuit of understanding the cosmos was a fundamental aspect of ancient societies, shaping their ideologies, rituals, and technological advancements. Remarkably, diverse civilizations developed profound interpretations of the universe, revealing a tapestry of celestial wisdom that transcended geographical boundaries and endured for millennia.

The cosmological beliefs of ancient civilizations were intertwined with their religious, mythological, and philosophical frameworks. In Mesopotamia, the Babylonians documented celestial omens on clay tablets, seeking to discern divine messages within celestial phenomena. Their astute observations laid the foundation for early astronomical calculations and the establishment of zodiacal constellations, reflecting the enduring legacy of their cosmic insights. Moreover, the Egyptians, revering the night sky as a reflection of the heavenly realm, aligned the construction of their pyramids with the positions of stars, demonstrating a profound connection between earth and cosmos.

Similarly, the ancient Greeks sought to rationalize the cosmos through philosophical inquiry, birthing foundational theories that transformed the course of Western thought. The concept of a geocentric universe, attributed to Ptolemy, echoed throughout the medieval era and profoundly influenced subsequent astronomical models. In contrast, the visionary heliocentric thesis proposed by Aristarchus of Samos challenged conventional perceptions of the cosmos, foreshadowing the revolutionary conceptions of Nicolaus Copernicus and Galileo Galilei.

Across eastern civilizations, the Chinese developed sophisticated cosmological models, integrating celestial observations with earthly events to navigate their agricultural calendars and dynastic rule. The harmonious interplay between celestial bodies and the human realm exemplified the integral relationship between cosmology and governance in ancient China, underscoring the practical applications of celestial knowledge.

Furthermore, the Mayans, renowned for their intricate calendrical systems and astronomical prowess, exhibited an acute understanding of celestial mechanics, evinced by their precise predictions of planetary movements and eclipses. Their cosmological pursuits were embedded within the fabric of daily life, guiding agricultural practices and religious ceremonies while resonating with a profound reverence for the interconnectedness of cosmic forces.

The cosmological insights of ancient civilizations resonate as enduring testaments to humanity's perennial quest for understanding the cosmos. By delving into the celestial mindsets of antiquity, we glean valuable insights into the origins of astronomical inquiry, the interplay of cosmology with societal frameworks, and the timeless allure of the celestial realms.

Astrological Instruments and Navigational Aids

Astrological instruments and navigational aids in ancient civilizations were not only marvels of advanced engineering but also essential tools for understanding the cosmos and navigating the vast expanses of land and sea. Ancient scholars and astronomers developed sophisticated devices to measure celestial movements, predict astronomical events, and aid in navigation across oceans and deserts. The astrolabe, an intricate instrument developed by Hellenistic astronomers, allowed for precise observations of celestial bodies, enabling accurate determination of latitude and time of day. This innovation revolutionized maritime navigation and greatly facilitated long-distance trade and exploration. Similarly, ancient Chinese astronomers designed the gnomon, a simple yet effective tool for measuring the sun's shadow, aiding in the determination of cardinal directions, the passage of time, and the changing seasons. This knowledge was crucial in agricultural planning and voyage preparations. Moreover, the Mayans, renowned for their advanced understanding of astronomy, constructed observatories with carefully aligned structures to track the movements of celestial bodies. They used these observations to develop sophisticated calendars, predict solar and lunar eclipses, and guide ritual practices. Across cultures, celestial maps and star charts provided valuable aids for navigation and astronomical study. Navigators relied on the positions of stars to determine direction and location during nighttime journeys, while astronomers used these maps to chart the paths of the planets and constellations. The rich tapestry of ancient astrological instruments and navigational aids demonstrates the universal human endeavor to comprehend the heavens and leverage this knowledge for practical applications. These innovations laid the foundation for modern astronomical advancements and continue to inspire contemporary explorations of the cosmos.

Ancient Observatories and Galactic Observations

Ancient civilizations across the world demonstrated an innate understanding of celestial events and their significance. The construction of observatories, such as the Mayan observatory at Chichen Itza or the astronomical complex at Chankillo in Peru, showcases the meticulous planning and precise architectural alignment focused on tracking celestial bodies. These ancient observatories were ingeniously designed to monitor the motions of stars, planets, and other heavenly phenomena. Predating the invention of the telescope, these structures harnessed the power of naked-eye observations to make accurate measurements and predictions related to astronomical events.

Furthermore, the astoundingly precise astronomical knowledge possessed by ancient cultures is evident in their calendrical systems and alignments of structures with celestial events. For instance, the alignment of the Great Pyramids of Giza with the constellation Orion's belt has been a subject of fascination and debate, hinting at the advanced astronomical understanding of the ancient Egyptians. The intricate patterns and alignments found in ancient observatories not only highlight the mathematical and engineering prowess of these societies but also indicate a deep spiritual and philosophical connection to the cosmos.

Among the most notable aspects of ancient observatories are the markings and inscriptions that provide insights into the interpretation of celestial activities. These records not only served as historical documentation but also contributed to the development of early astronomy and navigation. The knowledge gained from observing the night sky had practical applications in timekeeping, agricultural planning, and navigation, shaping the daily lives and cultural practices of these ancient civilizations.

Moreover, the study of the night sky held a profound symbolic and religious significance in many ancient societies. The alignment of key structures with solstices and equinoxes likely played a crucial role in religious rituals and ceremonies, emphasizing the interconnectedness of human existence with the cosmic cycles. Thus, the ancient observatories served as centers of both scientific inquiry and spiritual reverence, embodying the holistic approach that ancient cultures adopted in their pursuit of understanding the universe.

The legacy of ancient observatories and galactic observations continues to inspire contemporary scholars and scientists, providing valuable perspectives on humanity's enduring fascination with the mysteries of the cosmos. By delving into the methods and motivations behind these ancient endeavors, we gain profound insights into the intellectual curiosity and ingenuity of our forebears, fostering a deeper appreciation for the enduring legacy of ancient astronomical achievements.

The Antikythera Mechanism and Astronomical Predictions

The Antikythera mechanism, an extraordinary ancient artifact, exemplifies the sophistication of ancient technology in astronomy. Discovered in a shipwreck off the coast of the Greek island of Antikythera, this fascinating device dates back to the 2nd century BCE. Comprised of various cogwheels, dials, and inscriptions, the mechanism is believed to have been utilized for predicting celestial events and tracking astronomical cycles. Remarkably, it demonstrates an advanced understanding of complex mathematical and astronomical principles.

Given its intricate design, the Antikythera mechanism offers invaluable insights into how ancient civilizations comprehended the workings of the cosmos. It showcases their ability to calculate the positions of celestial bodies and forecast eclipses with remarkable precision, providing evidence of their remarkable achievements in observational astronomy and predictive modeling.

By unraveling the intricate intricacies of the Antikythera mechanism, contemporary scholars have gained profound respect for the technological prowess of ancient engineers and astronomers. Moreover, the artifact serves as a testament to the intellectual curiosity and innovation of our forebears, inspiring new generations to appreciate the depth of their knowledge and accomplishments in the field of astronomy.

Ongoing efforts to decipher the inscriptions on the mechanism and reconstruct its functionality continuously enhance our comprehension of ancient astronomical practices. Furthermore, interdisciplinary collaborations across the fields of archaeology, history, and astronomy continue to shed light on the implications of the Antikythera mechanism for our understanding of ancient astronomical practices and its potential influence on modern scientific endeavors.

As humanity embarks on further exploration of the cosmos, the Antikythera mechanism remains a poignant reminder of the enduring significance of ancient astronomical advancements. Its legacy continues to kindle admiration and reverence for the contributions of past civilizations to our collective understanding of the universe, serving as a bridge between the rich heritage of ancient astronomy and the ceaseless pursuit of new frontiers in space exploration.

Influence of Alien Theories on Contemporary Space Exploration Ethics

The notion of extraterrestrial life has captivated human imagination for centuries, and its influence on contemporary space exploration ethics is a topic of considerable significance. As we delve into the complex interplay between historical interpretations of ancient civilizations and modern theories about alien encounters, it becomes evident that these ideas have permeated societal values and ethical considerations in our quest to explore the cosmos. The allure of ancient artifacts and enigmatic constructions often fuels speculation about potential extraterrestrial influences, driving a fervent interest in uncovering connections between advanced technologies and otherworldly beings. This fascination has not only extended into popular culture but has also shaped ethical debates within the scientific community regarding the responsible pursuit of knowledge beyond our planet. It raises profound questions about how we should approach potential discoveries and interactions with non-terrestrial intelligence, emphasizing the need for a balanced and informed approach that respects both scientific integrity and cultural sensitivity. The impact of alien theories on contemporary space exploration ethics transcends mere speculation, influencing the development of protocols and guidelines for conducting research and exploration both within and beyond Earth's atmosphere. It compels us to reflect on the implications of these theories for our understanding of humanity's place in the universe and the ethical responsibilities that accompany our pursuits of cosmic discovery. Ultimately, the influence of alien theories serves as a thought-provoking lens through which we can critically assess our motivations, methodologies, and moral obligations as we venture into the great unknown of space.

Techniques of Construing Ancient Texts for Modern Applications

The translation and interpretation of ancient texts for modern applications is a multifaceted process that involves a deep understanding of historical context, linguistic nuance, and technological innovation. In the realm of space exploration, researchers and scholars have turned to ancient manuscripts, inscriptions, and artifacts to glean insights into the celestial knowledge and technological prowess of bygone civilizations. One of the fundamental techniques employed in this endeavor is philology, the study of language in written historical sources. By carefully analyzing the syntax, vocabulary, and cultural idioms present in ancient texts, philologists can unravel the meanings embedded in these often enigmatic writings. Furthermore, comparative analysis of multiple texts from different cultures can shed light on cross-cultural exchanges and shared astronomical knowledge. Another critical aspect of construing ancient texts lies in the integration of archaeological findings and material evidence. Inscriptions on celestial instruments, astronomical diagrams, and architectural alignments provide tangible evidence of ancient cosmological concepts and technological applications. By combining textual and material sources, researchers can reconstruct the intricate tapestry of ancient astronomical knowledge and devise methodologies for its application in contemporary space exploration. Moreover, interdisciplinary collaboration plays a pivotal role in this process, as experts in history, linguistics, archaeology, astronomy, and engineering converge to bridge the chasm between ancient wisdom and modern ingenuity. The utilization of computational tools and artificial intelligence algorithms has also revolutionized the analysis of ancient texts, enabling scholars to identify patterns, correlations, and predictive models embedded within cryptic writings. These cutting-edge techniques have unlocked new avenues for interpreting ancient cosmologies and leveraging them for space-related pursuits. As we embark on a journey to the stars, the nuanced amalgamation of traditional scholarship, technological innovation, and cross-disciplinary collaboration will continue to serve as the cornerstone for unraveling the secrets encoded in ancient texts and harnessing their relevance for contemporary space exploration endeavors.

Adaptation of Ancient Designs in Spacecraft Engineering

The study and adaptation of ancient designs for spacecraft engineering represent a convergence of historical knowledge, modern technology, and speculative innovation. By drawing inspiration from the technological ingenuity of ancient civilizations, contemporary spacecraft engineers have embraced an interdisciplinary approach to design and construction. One remarkable example of this trend is the incorporation of principles derived from ancient observatories and architectural marvels into the development of spacecraft navigation systems. By meticulously analyzing how ancient observers charted the movement of celestial bodies and constructed precise measurements, contemporary engineers have refined the accuracy and reliability of spacecraft guidance and positioning systems. Moreover, the integration of materials and structural concepts found in ancient architecture has revolutionized the construction and durability of spacecraft components. The resilience and longevity of ancient structures have informed the development of advanced composite materials and innovative design methodologies, ensuring the safety and endurance of modern space vehicles. In addition to these practical applications, spacecraft engineers have delved into interpretations of ancient texts and inscriptions, seeking potential blueprints for groundbreaking propulsion systems and energy sources. By synthesizing esoteric knowledge with cutting-edge scientific understanding, researchers are exploring the viability of conceptually reimagining ancient technologies to fuel and propel spacecraft beyond the confines of our solar system. The symbiosis between age-old wisdom and contemporary expertise has kindled a profound spirit of innovation, inspiring engineers to envision new frontiers in space exploration. The adaptability and resourcefulness exhibited by ancient civilizations serve as a testament to the enduring relevance of their achievements in informing and galvanizing the future of aerospace engineering. As the pursuit of space exploration continues, the legacy of ancient designs will undoubtedly continue to influence and shape the trajectory of humanity's quest for celestial discovery and transcendence.

Case Studies: Success Stories and Failures

In the exploration of ancient technology in space, numerous case studies shed light on both the triumphs and tribulations of integrating archaic designs into modern spacecraft engineering. These cases provide invaluable insights into the feasibility and challenges of reviving ancient concepts for contemporary space exploration initiatives. One such success story revolves around the incorporation of celestial navigation techniques from ancient civilizations, such as the Polynesians' mastery of non-instrumental navigational methods, which has inspired innovative approaches to spacecraft guidance systems. By harnessing these traditional methods, space missions have achieved remarkable accuracy and precision in interstellar travel, offering a testament to the efficacy of ancient practices in modern spacefaring endeavors. Conversely, examples of failures highlight the intricacies and limitations of implementing ancient technologies in space exploration. For instance, attempts to integrate ancient materials or structural designs into spacecraft construction have encountered significant hurdles related to compatibility, durability, and resilience in extraterrestrial environments. These setbacks underscore the necessity of meticulous research and rigorous testing when adapting archaic elements for contemporary aerospace applications. Despite these challenges, the comprehensive examination of both successful and unsuccessful ventures underscores the pivotal role of historical case studies in informing and enhancing the future utilization of ancient technology in space exploration. Through these diverse experiences, the boundary-pushing landscape of space exploration encapsulates a rich tapestry of achievements, setbacks, and paradigm-shifting discoveries, all of which contribute to the ongoing narrative of humanity's quest to combine the wisdom of ancient eras with the audacity of modern space exploration.

Future Prospective: Bridging Past Innovations with Tomorrow's Explorations

The future of space exploration is intrinsically linked to the past innovations that ancient civilizations have bestowed upon us. As we delve deeper into the mysteries of our universe, we are increasingly turning to the knowledge and achievements of our predecessors to inform and inspire our future endeavors. By bridging the gap between past and present, we not only honor the legacy of ancient ingenuity but also harness its potential to shape the course of space exploration.

Drawing from the expertise of ancient astronomers, engineers, and mathematicians, we can adapt their innovative concepts and insights to propel our modern pursuit of space exploration. The utilization of traditional astronomical instruments, such as astrolabes and celestial globes, offers valuable insights into navigation and celestial mapping that are relevant in contemporary space missions. Moreover, the intricate celestial calendars and predictive models developed by ancient cultures serve as a foundation for understanding cosmic cycles and optimizing mission timelines.

One of the most fascinating intersections between ancient technology and space exploration lies in the adaptation of ancient designs in spacecraft engineering. The principles of simplicity, durability, and precision evident in ancient mechanisms provide inspiration for the development of cutting-edge space technologies. By studying the craftsmanship of artifacts like the Antikythera Mechanism, we gain a profound appreciation for the advanced engineering capabilities of antiquity, which may offer innovative solutions for enhancing the functionality and efficiency of spacefaring vessels.

Furthermore, the deciphering of ancient texts and inscriptions holds great promise for aiding modern space exploration efforts. The process of interpreting ancient scripts not only enriches our understanding of historical celestial observations and phenomena but also unveils potential applications for contemporary space science. By leveraging linguistics, archaeoastronomy, and interdisciplinary methodologies, researchers can uncover hidden knowledge embedded within ancient writings, leading to groundbreaking advancements in space exploration strategies.

As we set our sights on the uncharted frontiers of space, it is imperative to consider the ethical implications and societal perspectives derived from ancient theories about extraterrestrial life. The exploration of alien encounters and cosmic consciousness in ancient myths and beliefs raises thought-provoking questions about humanity's place in the cosmos and the moral responsibilities associated with contact beyond Earth. By delving into these narratives, we cultivate a deeper understanding of our collective aspirations, fears, and aspirations, shaping the ethical frameworks and cultural dialogues relevant to contemporary space exploration initiatives.

In conclusion, as we chart a course toward the boundless realms of the universe, the wisdom and innovation of ancient civilizations serve as beacons lighting the way to our future in space exploration. By embracing the insights, tools, and philosophies of antiquity, we transcend temporal boundaries, embarking on a unified journey that integrates the wisdom of the past with the wonders of tomorrow.

26

Legacy of Ancient Technology in the Modern World

Historical Overview of Ancient Innovations

Archaeological findings and historical documents provide valuable insights into the evolution of ancient innovations, laying the foundation for modern technological methodologies. Delving into the annals of history, we unearth a rich tapestry of inventive techniques and revolutionary concepts that have transcended time and continue to influence contemporary civilization. The ingenuity of early civilizations, from Mesopotamia to Mesoamerica, is evident in their remarkable achievements in engineering, architecture, and other fields.

Ancient structures such as the Pyramids of Giza, the Great Wall of China, and the Parthenon stand as enduring testaments to the advanced knowledge and skills possessed by these ancient societies. Through meticulous examination of these monumental creations, we gain profound insights into the sophisticated engineering principles and architectural prowess that were employed thousands of years ago. These marvels not only showcase the technical capabilities of ancient artisans but also reflect their deep understanding of mathematics, astronomy, and practical problem-solving.

Moreover, the development of early technological methodologies can be discerned in the innovative tools and machines uncovered at archaeological sites. From intricate water systems in ancient Rome to the pioneering mechanical inventions in ancient Greece, these artifacts reveal the complex thinking and resourcefulness of our ancient predecessors. Insightful analysis of these relics illuminates the gradual progression of technological advancements, shedding light on the origins of fundamental engineering and mechanical concepts that continue to shape our world today.

The application of scientific principles in areas such as metallurgy, medicine, and urban planning further exemplifies the pioneering spirit of ancient innovators. Their empirical observations, experimental discoveries, and empirical practices laid the groundwork for myriad modern disciplines, setting enduring precedents for innovation and progress. By delving into the historical chronicles of these early technological methodologies, we come to appreciate the intellectual acumen and adaptive ingenuity of ancient civilizations, offering invaluable lessons and inspiration for contemporary pursuits of advancement and discovery.

Influences on Modern Engineering and Architecture

The architectural and engineering marvels of ancient civilizations continue to exert a profound influence on modern construction and design principles. From the awe-inspiring pyramids of Egypt to the grandeur of Roman infrastructure, ancient innovations have left an indelible mark on contemporary engineering and architecture. The ancient mastery of structural integrity, use of materials, and innovative construction techniques has not only stood the test of time but also serves as a source of inspiration for architects and engineers today. Through the study of ancient engineering feats, modern practitioners gain insights into efficient load-bearing designs, durable material utilization, and advanced water management systems. Furthermore, ancient architects developed sophisticated methods to manipulate light and space, creating spaces that evoke emotional and spiritual responses, a principle still valued in modern architectural practice. The enduring legacy of ancient technology is evident in the integration of sustainable design concepts drawn from historical practices. Applying ancient architectural principles, such as passive ventilation, natural lighting, and thermal massing, aligns with the contemporary emphasis on environmentally conscious and energy-efficient buildings. Additionally, the application of geometrical concepts and mathematical ratios employed by ancient builders continues to inform modern design aesthetics through the pursuit of harmony and balanced proportions. The global impact of ancient engineering extends to urban planning, as seen in the layout of ancient cities like Mohenjo-Daro and Athens, which influenced subsequent urban development models. The symbiotic relationship between ancient innovations and contemporary engineering and architecture underscores the enduring relevance of historical approaches and their capacity to enrich and inspire present-day designs. In essence, the legacy of ancient technology persists as a cornerstone of modern engineering and architecture, shaping our built environment and reflecting a timeless testament to human ingenuity.

Ancient Mathematical Principles in Contemporary Science

Throughout history, ancient civilizations have made remarkable contributions to mathematical principles that continue to influence contemporary scientific endeavors. The legacy of mathematical innovation can be traced back to the ancient Egyptians who developed sophisticated methods for surveying and constructing monumental structures. Their understanding of geometry and arithmetic laid the foundation for modern architectural and engineering practices. Similarly, the ancient Greeks, particularly mathematicians like Pythagoras and Euclid, formulated enduring geometric and algebraic theories that are fundamental to many scientific disciplines today. These principles have permeated fields such as physics, astronomy, and computer science, shaping the way we comprehend and interact with the world. Moreover, the Mesopotamians' advancements in numerical notation and calculation techniques have underpinned the development of complex algorithms and computational methodologies in contemporary science and technology. The legacy of Babylonian mathematics, including their base-60 numerical system and precise astronomical calculations, has left an indelible mark on modern quantitative analysis. The Indian subcontinent's mathematical heritage, as documented in texts like the 'Sulba Sutras,' introduced revolutionary concepts such as the concept of zero and the decimal numeral system, revolutionizing numerical representation and computation. These foundational contributions have been instrumental in shaping the intricate data analysis and modeling techniques employed across diverse scientific fields. Furthermore, the Chinese civilization's pivotal role in advancing mathematical knowledge, evident in their development of advanced arithmetic and algebraic ideas, has had a profound impact on contemporary applications of mathematics in areas such as cryptography, optimization, and operations research. By acknowledging and studying these ancient mathematical principles, the contemporary scientific community not only pays homage to the ingenious insights of our predecessors but also gains valuable perspectives and tools for addressing modern-day challenges. Embracing the enduring relevance of ancient mathematical knowledge enriches our collective understanding and empowers us to forge new frontiers in science and technology.

Legacy of Ancient Medicine and Pharmacology

The legacy of ancient medicine and pharmacology offers a captivating insight into the enduring influence of early healing practices on modern healthcare. Throughout history, civilizations such as those in ancient Egypt, Greece, China, and India developed sophisticated medicinal knowledge, blending empirical methods with spiritual beliefs and philosophical principles. The innovative treatments and remedies pioneered in these eras continue to resonate in contemporary medical practices.

Ancient medicinal traditions laid the groundwork for understanding the human body and the principles of disease. From herbal remedies and acupuncture to early surgical techniques, these practices fostered profound insights into anatomy, physiology, and pathology. For example, the use of medicinal plants in ancient cultures not only provided symptomatic relief but also contributed to the accumulation of knowledge that forms the basis of modern pharmacology. Furthermore, ancient medical texts, such as the Indian Ayurveda and the Greek Corpus Hippocraticum, document diverse therapeutic approaches and diagnostic methodologies that continue to inspire medical research today.

The legacy of ancient medicine extends beyond the realm of physical healing, encompassing the holistic well-being of individuals and communities. Ancient healers viewed health as a harmonious balance between the body, mind, and spirit, an approach that resonates with modern integrative medicine and holistic healthcare. Concepts of preventive medicine, dietetics, and wellness promotion, rooted in ancient wisdom, find resonance in contemporary public health initiatives and lifestyle medicine.

Moreover, the contributions of ancient medical practices to pharmacology are substantial. Many of the natural compounds and substances used in traditional medicines have yielded valuable sources for drug discovery and the development of modern pharmaceuticals. Archaeological findings have revealed the use of plant extracts and mineral compounds for therapeutic purposes in ancient civilizations, inspiring ongoing explorations of nature-based remedies and novel drug formulations.

In today's globalized world, the study of ancient medicine holds great significance in fostering cultural understanding and preserving indigenous healing knowledge. Collaborative efforts between modern medical practitioners and traditional healers aim to integrate the time-honored wisdom of ancient medicine with evidence-based healthcare, enhancing the accessibility and effectiveness of treatments. Through interdisciplinary research and cross-cultural dialogue, the legacy of ancient medicine continues to enrich the tapestry of modern healthcare, offering timeless lessons in compassion, resilience, and the pursuit of healing.

Environmental Strategies from the Ancients to Today

The environmental strategies employed by ancient civilizations continue to influence modern approaches to sustainable living and resource management. From the remarkably efficient water conservation techniques of the ancient Roman Empire to the intricate irrigation systems developed by civilizations along the Tigris and Euphrates rivers, ancient environmental practices provide valuable insights for contemporary challenges. The ancients' harmonious coexistence with nature, evident in their agricultural methods and city planning, serves as a testament to the enduring wisdom of these early societies. Modern environmentalists and urban planners have looked to these ancient strategies for inspiration, striving to replicate their sustainable principles in today's world. One notable example is the revival of ancient agricultural techniques such as terrace farming and agroforestry, which offer innovative solutions to current issues such as soil erosion and food security. Moreover, the communal water management systems established by ancient civilizations have influenced the development of modern water treatment and distribution networks, emphasizing the importance of collective responsibility for preserving and distributing this vital resource. Additionally, the principles of waste reduction and recycling observed in ancient cultures have inspired contemporary sustainability initiatives, encouraging the implementation of circular economies and responsible waste management practices. Furthermore, the ancient understanding of ecological balance and the interconnectedness of natural ecosystems has shaped the foundation of modern environmental science and biodiversity conservation efforts. By studying and adapting these timeless environmental strategies, society can progressively address pressing global concerns while respecting the ecological equilibrium that sustained our ancestors for centuries.

Impact on Modern Computing and Algorithms

The impact of ancient technologies on modern computing and algorithms is a fascinating study that reveals the enduring legacy of early innovations. Many foundational concepts in computing can be traced back to ancient civilizations, where rudimentary forms of data processing and logical reasoning were first conceived. In Ancient Egypt, for instance, sophisticated devices like the Antikythera mechanism demonstrated advanced mechanisms for representing and manipulating astronomical data, laying the groundwork for modern computational methods. Similarly, the Babylonians employed numerical systems and algorithms to solve complex mathematical problems, setting the stage for contemporary algorithmic thinking. The study of ancient civilizations has illuminated countless instances of proto-computational processes, inspiring modern scholars to explore the historical roots of computing. The development of modern computing languages owes much to the ancient world, with programming syntax often intersecting with ancient linguistic and symbolic systems. Furthermore, the principles of abstraction and algorithmic methodologies utilized by the ancients continue to inform contemporary software engineering practices. By understanding the ancient origins of computing, we gain valuable insights into the historical progression of computational thought and the enduring significance of early technological advancements. Moreover, the intersection of ancient metallurgy techniques and modern computing hardware underscores the link between ancient material science and contemporary technology. Advancements in metallurgical knowledge from antiquity have directly contributed to the development of semiconductors, nanomaterials, and other essential components of today's computational devices. The fusion of ancient material science with cutting-edge computing technology symbolizes the timeless nature of human innovation and its profound impact on shaping the modern world. As we continue to unravel the mysteries of ancient technologies, the integration of historical wisdom with evolving computational paradigms opens new frontiers for innovation and discovery. Ultimately, recognizing the influence of ancient computing and algorithms enriches our understanding of the dynamic relationship between the past and the present, offering valuable perspectives for charting the course of future technological evolution.

Ancient Metallurgy and Material Science in Modern Industry

The advancement of metallurgy and material science in ancient civilizations laid the foundation for modern industrial processes. The extraordinary achievements of our ancestors, from smelting ores to forging alloys, continue to influence and inspire contemporary manufacturing techniques. Ancient craftsmen honed their skills with remarkable precision, mastering the production of metals such as copper, bronze, iron, and steel. These early innovations revolutionized weaponry, tools, and infrastructure, providing a pivotal catalyst for technological progress.

In today's industrial landscape, the principles of ancient metallurgy remain intrinsic to the fabrication of complex alloys, structural materials, and cutting-edge composites. Engineers and scientists draw upon historical insights to optimize material properties for diverse applications, from aerospace components to medical implants. The in-depth understanding of phase transformations, heat treatment, and alloy design gleaned from ancient practices contributes to the development of high-performance materials with enhanced mechanical, thermal, and electrical characteristics.

Moreover, the sustainable ethos inherent in ancient metallurgical traditions continues to resonate in the modern industry. By embracing efficient utilization of resources and minimizing environmental impact, contemporary metalworking endeavors reflect a harmonious blend of ancestral wisdom and state-of-the-art innovation. Incorporating recycled materials, implementing energy-efficient processes, and adhering to eco-friendly standards underscore the enduring legacy of the ancients in promoting responsible industrial practices.

Furthermore, the preservation and restoration of ancient metal artifacts not only uphold cultural heritage but also provide invaluable insights for modern metallurgical research. Through non-destructive testing methods and advanced analytical techniques, scientists unravel the metallurgical secrets embedded in ancient relics, contributing to a deeper comprehension of material behavior under varying conditions.

As we navigate the complexities of material science and metallurgical engineering in the 21st century, it is imperative to recognize the profound impact of our forebears' expertise. Leveraging the enduring legacy of ancient metallurgy, contemporary industries stand poised to propel innovation, foster sustainability, and redefine the frontiers of material discovery and application.

Sustainability Lessons from Ancient Practices

Ancient civilizations offer an abundance of wisdom and knowledge that can be applied to contemporary sustainability efforts. The sustainable practices of ancient societies, born out of necessity and a deep respect for the environment, provide valuable lessons for current and future generations. One of the foundations of ancient sustainability was the harmonious integration of human activity with the natural world. From the irrigation systems of the Indus Valley civilization to the terraced agricultural landscapes of the Inca, ancient cultures demonstrated a profound understanding of ecological balance and resource management. These practices not only sustained their societies but also preserved the environment for centuries. The principles of conservation and resource efficiency that guided ancient civilizations are increasingly relevant in today's world. By studying the sustainable strategies of our ancestors, we gain insight into how to adapt our modern practices and technologies to minimize environmental impact. The concept of recycling and repurposing resources is deeply rooted in ancient traditions. For instance, the Mesoamerican practice of using terracotta pots for water storage and cooling dates back thousands of years and is a testament to their innovative use of locally available materials. Similarly, the Roman Empire's sophisticated aqueduct system showcases their mastery of hydraulic engineering, enabling efficient water distribution and promoting sustainable living. Moreover, the indigenous peoples of North America practiced sustainable land management, using controlled burning to maintain healthy ecosystems and promote biodiversity. Their holistic approach to stewardship of the land offers valuable lessons for mitigating the impact of deforestation and industrial agriculture on the environment. As we confront global challenges such as climate change and resource depletion, embracing the sustainability lessons from ancient cultures becomes imperative. Learning from their time-honored practices reminds us that sustainable living is not a new concept but a timeless ethos deeply woven into the fabric of human history. Embracing these lessons involves re-evaluating our relationship with nature and restructuring our societal frameworks to prioritize long-term ecological harmony over short-term gains. By applying the sustainable strategies of our ancestors, we can cultivate a more sustainable future for all life on Earth.

Education and Knowledge Transmission Across Ages

Education and knowledge transmission have been fundamental elements of human society since ancient times. In the context of ancient technology, the transfer of knowledge was primarily achieved through oral traditions, apprenticeships, and master-disciple relationships. The passing down of skills and expertise from one generation to another formed the backbone of technological progress in civilizations such as Mesopotamia, Ancient Egypt, China, Greece, and Rome.

These early societies recognized the value of preserving and transferring knowledge, paving the way for educational systems that evolved over time. From the establishment of ancient academies and schools to the pioneering contributions of scholars and polymaths, the dissemination of knowledge became formalized, expanding access and fostering innovation. The great libraries of antiquity, such as the Library of Alexandria in Egypt, served as centers of learning and repositories of vast knowledge, making valuable information accessible to a wider audience.

As civilizations transitioned into the medieval and renaissance periods, the proliferation of universities and centers of learning gave rise to more structured forms of education. This facilitated the sharing of knowledge across various disciplines, from philosophy and astronomy to mathematics and medicine. The legacy of these educational institutions continues to shape modern academia, with their influence permeating fields ranging from humanities and social sciences to STEM disciplines.

In the contemporary era, technological advancements have revolutionized the landscape of education and knowledge transmission. Access to information has become democratized through digital platforms and e-learning, enabling individuals around the globe to engage with diverse sources of knowledge. The integration of ancient wisdom with innovative technologies has led to the development of immersive learning experiences, interactive tools, and virtual simulations, transcending barriers of time and space.

Furthermore, interdisciplinary studies and collaborative research initiatives reflect the interconnectedness of knowledge transmission across ages. By synthesizing ancient teachings with contemporary discoveries, this approach fosters holistic understanding and promotes cross-cultural exchange. Acknowledging the enduring relevance of traditional practices, educational curricula increasingly incorporate elements of ancient wisdom, nurturing a balanced and comprehensive learning environment.

In summary, the evolution of education and knowledge transmission across ages encapsulates the continuous quest for intellectual enrichment and scholarly advancement. Drawing from the roots of ancient wisdom while embracing the frontiers of modern innovation, this seamless interplay propels the collective pursuit of knowledge towards new horizons, enriching both individual minds and global discourse.

Future Prospects: Integrating Ancient Wisdom with Innovative Technologies

The future holds immense promise for the integration of ancient wisdom with innovative technologies. As humanity continues to advance in various fields, there is a growing recognition of the value that ancient knowledge and practices offer in shaping a sustainable and prosperous future. By leveraging the timeless wisdom of our ancestors and combining it with cutting-edge technologies, we can pave the way for unprecedented progress across multiple domains.

In the realm of sustainable development, ancient wisdom offers invaluable insights into harmonious coexistence with nature. Lessons from indigenous cultures and historical civilizations provide a lens through which we can reimagine our relationship with the environment. By integrating ancient agricultural techniques, resource management practices, and ecological philosophies with modern innovations such as precision farming, renewable energy, and eco-friendly infrastructure, we can create a more balanced and sustainable world for future generations.

Moreover, the convergence of ancient knowledge with innovative technologies holds tremendous potential in the healthcare sector. Traditional healing systems and herbal remedies passed down through generations can be synergized with advanced medical research and biotechnology to develop new treatments, enhance holistic wellness approaches, and address complex health challenges. Furthermore, incorporating ancient mindfulness practices and mental well-being techniques into modern mental healthcare can enrich therapeutic interventions and promote overall psychological resilience.

The fusion of ancient wisdom with innovative technologies extends to the realms of architecture and urban planning as well. By drawing inspiration from the architectural marvels of antiquity and infusing them with state-of-the-art sustainable design principles, urban areas can undergo transformational developments that prioritize both aesthetic appeal and environmental responsibility. Strategies such as passive building design, natural ventilation systems, and adaptive reuse of traditional construction methods can create truly regenerative urban spaces that honor the past while embracing the future.

In the ever-evolving landscape of education, the integration of ancient pedagogical approaches with emerging educational technologies stands to revolutionize learning experiences. By incorporating time-tested teaching methodologies, knowledge stewardship practices, and interdisciplinary learning frameworks, educators can foster a more profound appreciation for diverse cultural heritages while equipping students with the critical thinking and adaptability needed to navigate the complexities of the modern world.

Furthermore, as advancements in artificial intelligence, machine learning, and data analytics continue to redefine industries, the ethical and moral foundations embedded in ancient philosophical traditions can serve as guiding beacons in the development and application of these technologies. By integrating ethical considerations derived from ancient wisdom into the fabric of technological innovation, we can build a future where humanistic values and technological progress coexist harmoniously, mitigating potential risks and ensuring that innovation serves the collective good.

Ultimately, the prospects of integrating ancient wisdom with innovative technologies are boundless. As we stand on the precipice of a new era defined by rapid technological advancement, embracing the enduring wisdom of our ancestors provides us with a compass to navigate the uncharted territories of the future. By heeding the lessons of antiquity and synergizing them with cutting-edge innovations, we can forge a path towards a more enlightened, sustainable, and harmonious tomorrow.

27

Ancient Engineering in the Americas

The ancient civilizations of the Americas were renowned for their remarkable engineering feats, demonstrating advanced knowledge in architecture, irrigation, and infrastructure. From the impressive city of Teotihuacan in Mesoamerica to the enigmatic stone monuments of the Andes, the indigenous peoples of the Americas displayed a sophisticated understanding of engineering principles that continues to captivate modern researchers.

In Mesoamerica, the city of Teotihuacan, located in present-day Mexico, is a dazzling example of ancient urban planning and engineering. The central thoroughfare, known as the Avenue of the Dead, is aligned with precision to the solstices, showcasing a profound understanding of celestial movements. The grand pyramids of the Sun and Moon, towering over the city, stand as enduring testaments to the engineering prowess of the Teotihuacanos. These massive structures, constructed without the use of metal tools or beasts of burden, remain a testament to the ingenuity and organizational capacity of this ancient civilization.

The Inca civilization, centered in the Andes mountains of South America, is renowned for its remarkable feats of engineering. The Inca constructed vast networks of roads, some of which traversed treacherous mountain terrain at elevations exceeding 13,000 feet. Additionally, the Inca's sophisticated agricultural terraces, known as "andenes," showcased their mastery of hydraulic engineering and soil conservation techniques. Their terraced farming system allowed for efficient land use and enabled cultivation at diverse altitudes, proving the Inca's deep understanding of environmental and geological considerations.

Moreover, the ancient engineers of the Americas displayed remarkable proficiency in constructing complex irrigation systems. The Nazca civilization, known for its enigmatic Nazca Lines in Peru, demonstrated a sophisticated understanding of aqueducts and underground channels, enabling the efficient distribution of water in the arid coastal desert. The complexity and precision of their water management systems highlight the ingenuity and foresight of the Nazca people. Similarly, the Hohokam people in present-day Arizona constructed an extensive network of canals for agricultural irrigation, showcasing their adeptness in hydrology and hydraulic engineering. These canals were integral to the agricultural productivity of the region and demonstrated the Hohokam's understanding of sustainable water management.

In addition to monumental structures and irrigation systems, the ancient engineers of the Americas also displayed remarkable precision in their stonework. The ancient city of Machu Picchu in Peru, constructed by the Inca, features meticulously crafted stone structures that have withstood centuries of environmental challenges. The sophisticated use of stone masonry and precise fitting of stones in the construction of walls, buildings, and temples at Machu Picchu stand as a testament to the Inca's mastery of architectural engineering.

The impressive structures and engineering marvels of the Americas reveal the advanced knowledge and ingenuity of ancient civilizations in this region. The enduring legacy of their engineering feats continues to evoke admiration and curiosity, inspiring contemporary researchers and engineers to unravel the secrets of these ancient engineering accomplishments.

The study of ancient engineering in the Americas provides valuable insights into the societal organization, technological advancements, and cultural values of these pre-Columbian civilizations. The intricate understanding and application of astronomy, mathematics, and geology by ancient American engineers paved the way for innovative urban planning, sustainable resource management, and monumental construction projects that continue to fascinate and inspire scholars and architects. Moreover, the intricate interplay between environmental adaptation and engineering solutions, as seen in the terraced agriculture of the Inca and the water management systems of the Hohokam and Nazca, exemplifies an advanced ecological awareness and utilization of natural resources.

Furthermore, the impressive scale of the ancient American engineering achievements highlights the remarkable societal organization and labor mobilization of these cultures. The construction of monumental pyramids, road networks, and intricate water management systems required meticulous planning, expert craftsmanship, and a deep understanding of the local topography. This underscores the intricate social and political structures that facilitated such large-scale collaborative endeavors, providing valuable insights into the governance and labor dynamics of these ancient societies.

In conclusion, the legacy of ancient engineering in the Americas is a testament to the enduring impact of indigenous knowledge and ingenuity. The complex engineering feats and architectural wonders of ancient American civilizations offer a window into the rich tapestry of human innovation and adaptation to diverse environments. By delving into the achievements of these ancient engineers, we gain a deeper appreciation for the intellectual and technical prowess of pre-Columbian societies, while also recognizing the enduring influence of their accomplishments on contemporary engineering practices and urban planning. The remarkable engineering feats of the ancient Americas continue to inspire awe and admiration, prompting continuous study and exploration. As researchers delve deeper into these ancient marvels, new discoveries emerge, shedding light on the intricate techniques and tools employed by ancient engineers.

Excavations and surveys at Teotihuacan have revealed a complex system of subterranean tunnels and chambers beneath the city, hinting at the possibility of intricate hydraulic and drainage systems that contributed to the sustainability and functionality of the urban center. These discoveries offer a deeper understanding of the engineering ingenuity of the Teotihuacanos and the sophisticated infrastructure that underpinned their monumental city.

In the Andes, ongoing research into the Inca road network has unveiled the complexity and precision of their construction methods. Utilizing a system of cut-stone blocks and retaining walls, the Inca engineered roads that seamlessly integrated with the rugged mountain terrain, presenting a remarkable blend of natural and human-made landscapes. Furthermore, recent studies suggest that the Inca roads also served as sophisticated communication networks, allowing for the efficient dissemination of information and goods across their vast empire.

In the coastal desert region of Nazca, investigations into the water management systems have uncovered a network of canals and reservoirs, demonstrating the Nazca's adeptness at harnessing scarce water resources for agricultural purposes. The mastery of hydrology and the understanding of water flow dynamics exhibited by the Nazca highlight their advanced knowledge in engineering and resource management.

Similarly, the Hohokam canal system in present-day Arizona continues to be a subject of extensive research, revealing the intricate network of channels, check dams, and reservoirs that sustained agricultural productivity in the arid landscape. Furthermore, studies of the Hohokam canal construction techniques and maintenance practices provide valuable insights into the sustainable use of water resources and the integration of engineering knowledge with environmental adaptation.

The enduring mystery and allure of ancient American engineering marvels continue to drive ongoing archaeological and interdisciplinary research. Advanced technologies such as LiDAR (Light Detection and

Ranging) and ground-penetrating radar are revolutionizing the study of ancient landscapes, uncovering hidden structures and infrastructure that were previously inaccessible to conventional archaeological methods.

Moreover, interdisciplinary collaborations between archaeologists, geologists, climatologists, and engineers are shedding new light on the intricate relationships between ancient engineering, environmental factors, and societal development. By analyzing geological formations, water flow patterns, and climatic data, researchers are gaining a holistic understanding of how ancient civilizations adapted to and modified their environments through engineering interventions.

The study of ancient engineering in the Americas offers a profound appreciation for the creativity, resilience, and adaptability of the indigenous peoples who crafted these enduring marvels. The legacy of their engineering accomplishments continues to inspire contemporary efforts in sustainability, infrastructure development, and cultural preservation, highlighting the enduring relevance of ancient wisdom in modern times. As researchers continue to unravel the engineering secrets of the Americas, they pave the way for innovative approaches to contemporary challenges, drawing from the rich reservoir of knowledge and practices developed by ancient civilizations.

One of the most exciting aspects of ongoing research in ancient American engineering is the potential to engage with indigenous communities and incorporate their traditional knowledge and practices into the study of ancient engineering marvels. Collaborative efforts with local communities can provide invaluable insights into the cultural significance, oral traditions, and practical applications of ancient engineering techniques, ensuring a more holistic and culturally sensitive interpretation and preservation of these extraordinary legacies.

Furthermore, the study of ancient American engineering opens up opportunities for educational outreach and public engagement. Museums, archaeological sites, and educational institutions can use the allure of ancient engineering marvels to inspire interest in science, technology, engineering, and mathematics (STEM) fields, fostering a deeper appreciation for the ingenuity and achievements of past civilizations.

In a global context, the study of ancient American engineering serves as a powerful reminder of the diversity of human ingenuity and the multitude of ways in which different cultures have harnessed their environments to create sustainable and resilient societies. By celebrating the accomplishments of ancient engineers in the Americas, we honor the enduring legacy of indigenous knowledge and reaffirm the importance of cultural diversity in shaping our understanding of the past and guiding our actions for the future.

As researchers continue to unravel the mysteries of ancient American engineering, they hold a responsibility to ensure that their work promotes cultural respect, ethical engagement, and collaborative partnerships with indigenous communities. By centering the voices and perspectives of descendant communities, archaeologists and engineers can contribute to a more inclusive and equitable representation of the ancient Americas' engineering achievements, fostering mutual respect and understanding across cultural boundaries.

In the spirit of honoring the engineering prowess of ancient American civilizations, contemporary engineers and researchers can draw inspiration from these ancient marvels to address pressing global challenges in sustainable infrastructure, water conservation, and ecological resilience. By embracing the wisdom of the past and integrating it with modern innovations, we can strive for a more harmonious relationship with our environment and pave the way for a future that respects and preserves the diverse knowledge systems that have shaped human history.

In conclusion, the study of ancient engineering in the Americas serves as a testament to the remarkable ingenuity, resilience, and cultural richness of indigenous societies. By unveiling the secrets of these ancient marvels, we not only gain insights into the technological achievements of the past but also reaffirm the enduring relevance of ancient wisdom in informing our actions and decisions in the present day. As we continue to explore the engineering feats of the ancient

Americas, we embark on a journey of discovery, appreciation, and mutual learning that bridges the past and the present, honoring the extraordinary achievements of those who came before us.

www.ingramcontent.com/pod-product-compliance
Lightning Source LLC
Chambersburg PA
CBHW072147070526
44585CB00015B/1026